T5-ARC-432

167

Advances in Polymer Science

Advances in Polymer Science

Recently Published and Forthcoming Volumes

Polymer Synthesis · Polymer Analysis
Vol. 171, 2004

NMR · Coordination Polymerization · Photopolymerization
Vol. 170, 2004

Long-Term Properties of Polyolefins
Volume Editor: Albertsson, A.-C.
Vol. 169, 2004

Polymers and Light
Volume Editor: Lippert, T.
Vol. 168, 2004

New Synthetic Methods
Vol. 167, 2004

Polyelectrolytes with Defined Molecular Architecture II
Volume Editor: Schmidt, M.
Vol. 166, 2004

Polyelectrolytes with Defined Molecular Architecture I
Volume Editor: Schmidt, M.
Vol. 165, 2004

Filler-Reinforced Elastomers · Scanning Force Microscopy
Vol. 164, 2003

Liquid Chromatography · FTIR Microspectroscopy · Microwave Assisted Synthesis
Vol. 163, 2003

Radiation Effects on Polymers for Biological Use
Volume Editor: Kausch, H.
Vol. 162, 2003

Polymers for Photonics Applications II
Nonlinear Optical, Photorefractive and Two-Photon Absorption Polymers
Volume Editor: Lee, K.-S.
Vol. 161, 2003

Filled Elastomers · Drug Delivery Systems
Vol. 160, 2002

Statistical, Gradient, Block and Graft Copolymers by Controlled/ Living Radical Polymerizations
Authors: Davis, K.A., Matyjaszewski, K.
Vol. 159, 2002

Polymers for Photonics Applications I
Nonlinear Optical and Electroluminescence Polymers
Volume Editor: Lee, K.-S.
Vol. 158, 2002

Degradable Aliphatic Polyesters
Volume Editor: Albertsson, A.-C.
Vol. 157, 2001

Molecular Simulation · Fracture · Gel Theory
Vol. 156, 2001

New Polymerization Techniques and Synthetic Methodologies
Vol. 155, 2001

Polymer Physics and Engineering
Vol. 154, 2001

New Synthetic Methods

With contributions by
Y. Chujo · R. Faust · Y. Kwon · K. Naka
J.J. Robin · T. Uemura · O.W. Webster

The series presents critical reviews of the present and future trends in polymer and biopolymer science including chemistry, physical chemistry, physics and material science. It is addressed to all scientists at universities and in industry who wish to keep abreast of advances in the topics covered.

As a rule, contributions are specially commissioned. The editors and publishers will, however, always be pleased to receive suggestions and supplementary information. Papers are accepted for "Advances in Polymer Science" in English.

In references Advances in Polymer Science is abbreviated Adv Polym Sci and is cited as a journal.

The electronic content of APS may be found at http://www.springerLink.com

ISSN 0065-3195
ISBN 3-540-00544-7
DOI 10.1007/b10950
Springer-Verlag Berlin Heidelberg New York

Library of Congress Catalog Card Number 61642

Springer-Verlag is a part of Springer Science+Business Media

springeronline.com

Typesetting: Stürtz AG, Würzburg
Cover: Künkellopka GmbH, Heidelberg; design&production GmbH, Heidelberg

Printed on acid-free paper 02/3020/kk - 5 4 3 2 1 0

Editorial Board

Advances in Polymer Science
Also Available Electronically

For all customers who have a subscription to Advances in Polymer Science, we offer the electronic version via SpringerLink free of charge. Please contact your librarian who can receive a password for free access to the full articles by registering at:

http://www.springerlink.com

If you do not have a subscription, you can still view the tables of contents of the volumes and the abstract of each article by going to the SpringerLink Homepage, clicking on "Browse by Online Libraries", then "Chemical Sciences", and finally choose Advances in Polymer Science.

You will find information about the

- Editorial Board
- Aims and Scope
- Instructions for Authors
- Sample Contribution

at http://www.springeronline.com using the search function.

Contents

MMA	Methyl methacrylate
MTS	1-Methoxy-1-(trimethylsiloxy)-2-methylprop-1-ene
Mn	Number average molecular weight
Mw	Weight average molecular weight
MWD	Molecular weight dispersity, Mw/Mn
MW	Molecular weight
P	Polymer
P5	Tetrakis[tris(dimethylamino)phosphoranylidenamino]phosphonium
PBMA	Poly(butyl methacrylate)
PMA	Poly(methyl acrylate)
PMMA	Poly(methyl methacrylate)
RAFT	Reversible addition fragmentation transfer
S	Styrene
SFRP	Stable free radical polymerization
TAS	Trisdimethylaminosulfonium
TBA	Tetrabutylammonium
tol	Tolyl
TEMPO	2,2,6,6-Tetramethyl-1-piperidinyloxy
TMS	Trimethylsilyl

1 Introduction

Twenty years have passed since DuPont announced a startling new process for polymerization of methacrylate monomers [1]. The method uses a trimethylsilyl ketene acetal initiator catalyzed by nucleophilic anions. It operates at 80 °C and gives unprecedented control over polymer chain architecture (Scheme 1).

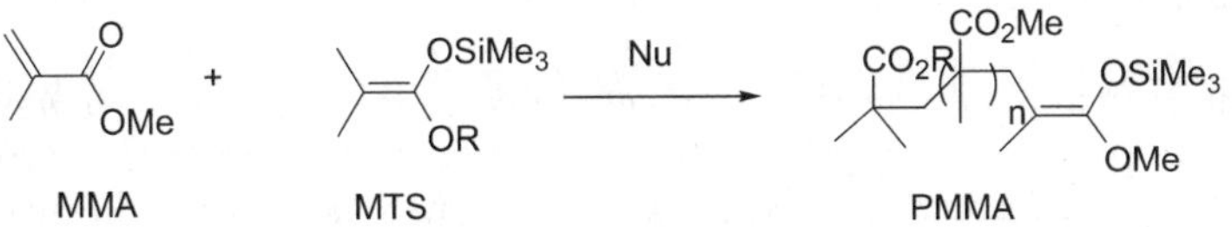

MMA + MTS (OSiMe$_3$, OR) $\xrightarrow{Nu}$ PMMA (CO_2R, CO_2Me, n, $OSiMe_3$, OMe)

Scheme 1

Based on evidence available at the time, the DuPont workers proposed that the trimethylsilyl group was transferring to monomer as it was adding to the polymer chain ends and thus named the new procedure Group Transfer Polymerization (GTP). Based on all the evidence *now* available this mechanism is almost certainly wrong but the name should remain since it is firmly imbedded in the chemical literature.

There are already several excellent detailed reviews on GTP [2–5]. In this chapter I will critically analyze the existing data that strongly support a dissociative (anionic) mechanism originally proposed by R. Quirk of Akron University [6]. I will also explain how GTP can operate at 80 °C when it is well known that the classical anionic polymerization of methacrylates does not proceed above ambient temperatures. In addition, GTP will be compared to other controlled polymerization methods.

2 Desirable Attributes for Commercial Controlled Polymerization of (Meth)acrylates

During the mid-1970s DuPont investigated the living polymerization of methacrylate monomers by anionic initiation at –80 °C [7]. Di and tri block polymers were made that had potential for use as pigment dispersing agents and for rheology control. The project was abandoned when calculations showed that the refrigeration necessary to keep the reactors at –80 °C was too costly. To be commercially viable the following characteristics for controlled polymerization were deemed necessary:

- Operating temperatures in the 50–80 °C range so that reflux condensers can be used to remove the heat of polymerization
- The ability to make block polymers containing no more than 10% homopolymer (minimal chain termination)
- Nearly colorless product
- Low sensitivity to impurities
- Overall cost of resin under $5/lb
- Ability to produce resin with molecular weights in the 20,000 range and molecular weight distributions (MWDs) under 1.3
- Minimal metallic or halide impurities in the final product
- Nontoxic ingredients
- Product free from unpleasant odors

Group transfer polymerization meets most of these criteria. However, it is sensitive to protic impurities and the present cost of the initiator is too high. Other living processes for polymerization of (meth)acrylates will be evaluated with respect to these criteria.

3 The GTP Process

As shown in Scheme 1, GTP converts methacrylate monomer to a polymer with one end group corresponding to the R on the initiator and the other end a trimethylsilyl ketene acetal. If the initiator contains a vinyl group not reactive to GTP, a macromonomer results [8]. The silyl ketene acetal end can be used to initiate another monomer, for example butyl methacrylate, to give a block polymer (Scheme 2b) [1] or with other reagents to add additional functionality. Thus benzaldehyde generates a phenylhydroxymethyl end [9]. Reaction of silyl ended polymer with a difunctional methacrylate (Scheme 2c) generates a multi-armed star polymer (Scheme 2d) [10]. GTP can be quenched by addition of methanol (Scheme 2e) [1]. Thus one can safely abort a polymerization run that is proceeding too fast.

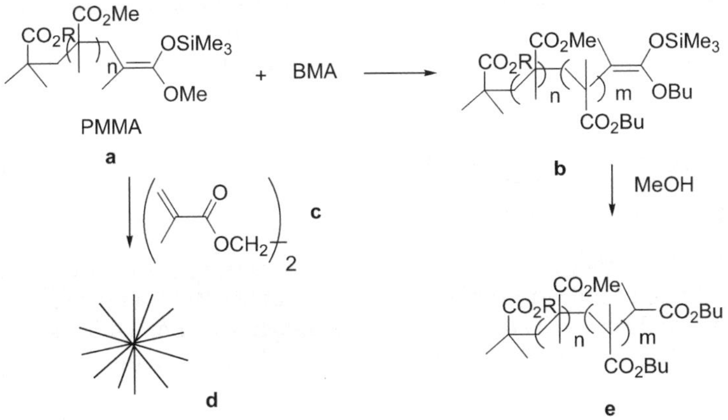

Scheme 2

GTP operates at temperatures up to 100 °C when catalyzed by weak nucleophiles such as tetrabutylammonium benzoate [11]. Molecular weights in the 20,000 range are easily obtained but generation of polymer in the 60,000 range is difficult. As in other living systems the molecular weight is controlled by the monomer/initiator ratio and the MWDs are narrow. During the polymerization especially at higher temperatures the resulting polymer will contain up to 30% dead ends, the result of backbiting of enolate chain ends (Scheme 3) and/or reaction with protic impurities [10]. Brittain [12] has studied chain termination and found that at the trimer stage (3 DP) the rate of backbiting is ten times that occurring at higher DPs. Thus one should start GTP with more than three equivalents of monomer to by-pass the trimer stage quickly.

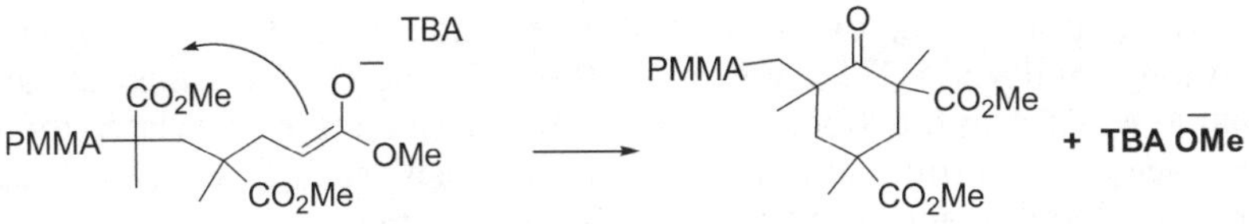

Scheme 3

The tetrabutylammonium (TBA) catalysts are slowly consumed during the progress of the polymerization and in longer runs additional amounts must be added with the monomer to keep the polymerization going. Since the TBA cation, if it still exists, must be associated with an anion, it follows that the catalytic activity would not decrease unless the TBA ion had been consumed. Reetz has shown that tributylamine is produced during anionic polymerization with TBA as the counterion, no doubt the result of Hoffman elimination or SN2 attack on the TBA [13] (Scheme 4). Butene was detected but no quantitative measurement was made.

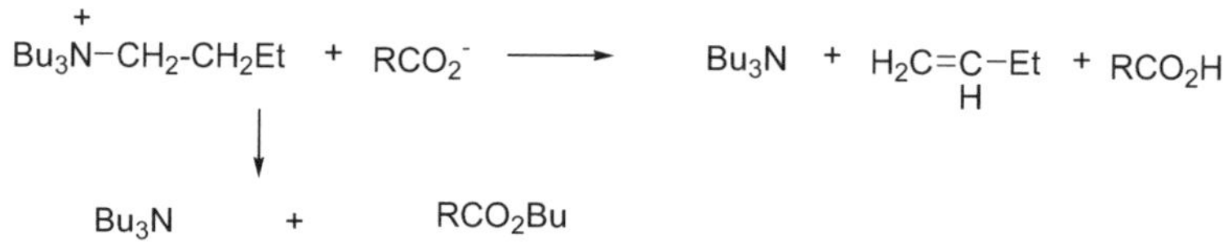

$$Bu_3\overset{+}{N}-CH_2-CH_2Et + RCO_2^- \longrightarrow Bu_3N + H_2C{=}\underset{H}{C}-Et + RCO_2H$$

$$\downarrow$$

$$Bu_3N + RCO_2Bu$$

Scheme 4

It would be of interest to see if the more stable cesium carboxylates would be better catalysts for GTP than TBA carboxylates since Quirk showed that cesium 9-methylfluorenide works as well as TBA 9-methylfluorenide as a catalyst in his mechanism studies [6].

3.1 GTP Monomers

By far the best monomers for GTP are the methacrylates. Glycidyl methacrylate and other substituted members of the family can be used to make highly functional block polymers. If the monomer contains active hydrogen, for example, hydroxyethyl methacrylate or methacrylic acid, GTP does not proceed. These functions can, however, be masked by trimethylsilyl groups [9] (Scheme 5).

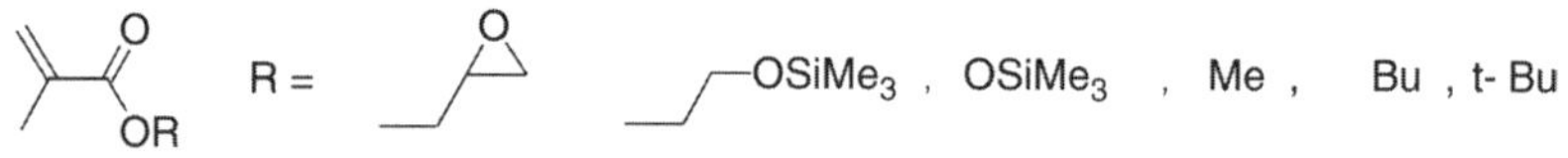

O, OR, R = , OSiMe₃ , OSiMe₃ , Me , Bu , t- Bu

Scheme 5

Acrylates polymerize two orders of magnitude faster than methacrylates by anion catalyzed GTP; however, the polymerization dies at about 10,000 MW. During the anion catalyzed polymerization of acrylates the silyl ketene acetal end groups migrate to internal positions. These ketene acetals are too hindered to act as initiators for branch formation [9].

The living polymerization of acrylates by GTP does proceed under Lewis acid catalysis [14]. $ZnCl_2$ or $ZnBr_2$ are effective but require concentrations of catalyst at a level of 10% based on monomer. R_2AlCl works at lower levels. However, $HgCl_2$ activated by TMS iodide is the best Lewis acid system and gives living acrylate polymers at low catalyst levels [15]. For industrial use the toxicity of mercury compounds is a problem.

Under GTP conditions conjugated dienoates as well as trienoates polymerize faster than methacrylates with anionic catalysis. The dienesilyl acetal **6a** is a better initiator than MTS [16] (Scheme 6).

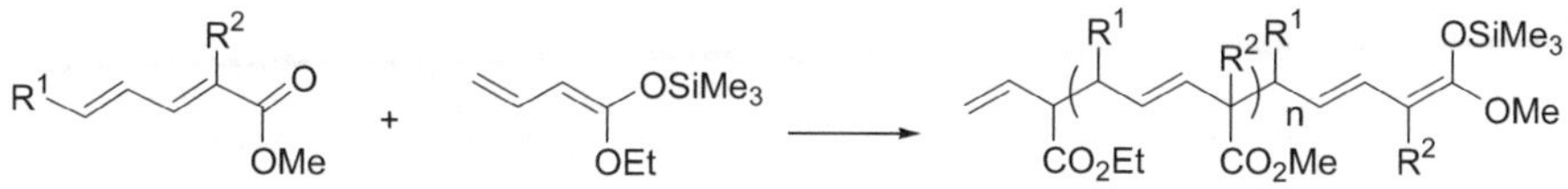

R^1 = Me or H
R^2 = Me or H

Scheme 6

Acrylonitrile and methacrylonitrile (MAN) polymerize extremely rapidly by GTP [9]. Molecular weight control is difficult since the polymers form before uniform mixing occurs. Although **7a** can be used to initiate MAN, MTS is better. The silyl imine end **7b** is the likely chain carrier (Scheme 7).

Scheme 7

3.2 Aldol GTP

In a process related to GTP, aldehydes initiate the polymerization of silyl vinyl ethers and silyl diene ethers. Here the silyl group is present in the monomer and transfers to the aldehyde ended chains regenerating aldehyde ends [17] (Scheme 8). A Lewis acid catalyst is required. *tert*-Butyldimethylsilyl works best as a transfer group for vinyl ether while trimethylsilyl is suitable for diene ethers [18]. Even though aldol GTP provides a route to polyvinyl alcohol segments in the subsequent block polymer synthesis, the projected cost of the monomers discouraged further research aimed at commercialization.

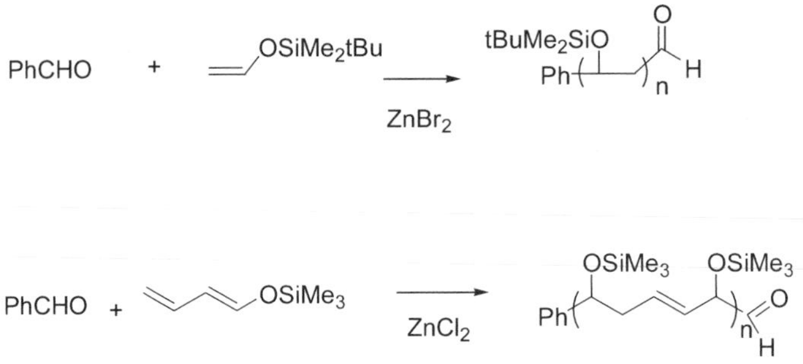

$$PhCHO + CH_2{=}CH{-}OSiMe_2tBu \xrightarrow{ZnBr_2} Ph{-}[CH(OSiMe_2tBu){-}CH_2]_n{-}CHO$$

$$PhCHO + CH_2{=}CH{-}CH{=}CH{-}OSiMe_3 \xrightarrow{ZnCl_2} Ph{-}[CH(OSiMe_3){-}CH_2{-}CH{=}CH]_n{-}CHO$$

Scheme 8

3.3 Initiators for GTP

To obtain polymer with low MWDs in a living polymerization the rate of initiation must be faster or similar to the rate of propagation. This can suitably be accomplished if the structure of the initiator is the same as that of the growing chain end. For GTP this is a silyl ketene acetal (Scheme 9). The

a: $R^1R^2C{=}C(SiR^3_3)(OR^4)$ b: Me_3SiCN c: $(RO)_2POSiMe_3$ d: Me_3SiSR

If $R^1 = R^2 = R^3 = R^4 = Me$ then "a" is MTS

Scheme 9

pre-ferred initiator for GTP is MTS where R^1, R^2, R^3, and R^4 are all methyl groups [1]. MTS is made by addition of trimethylsilane to MMA and, although it is commercially available, it is relatively expensive. A low cost initiator would greatly increase the use of GTP for commercial products. If R^1 is large and R^2 methyl, initiation is not retarded but if both R^1 and R^2 are large the ketene acetal will not work as an initiator [9]. Trimethylsilyl is ideal for the silyl function. Phenyldimethylsilyl seems to have about the same activity as MTS but *tert*-butyldimethyl [9] and triethylsilyl [19] groups are much less reactive. Large groups on the ether function R^4 do not appear to affect the rate of initiation [9]. Silyl ketone enolates initiate GTP but the rate of initiation is slower than the rate of propagation [20]. A number of silicon compounds initiate GTP by first adding to the methacrylate monomer to generate silyl ketene acetals (Scheme 9b,c,d).

3.4 Catalysts for GTP

3.4.1 *Nucleophilic Anions*

The preferred catalysts for GTP are nucleophilic anions. The most active catalysts are fluorides and bifluorides [1]. At above ambient temperatures, however, carboxylates and bicarboxlates are preferred [11]. A large counter ion is required for maximum efficiency. In the early work trisdimethylaminosulfonium (TAS) was used, but later the more readily available tetrabutylammonium (TBA) salts have gained favor. Since TBA slowly decomposes under the basic conditions used for GTP, other positive ions may work better. Quirk used cesium ion for his mechanistic studies and found it to be equivalent to TBA [6]. Bywater worked with the very stable $Ph_3PNPPh_3^+$ bifluoride in his mechanistic probes [19] and Jenkins [21] showed that potassium complexed with 18-crown-6 was a possible alternative to TBA (Scheme 10).

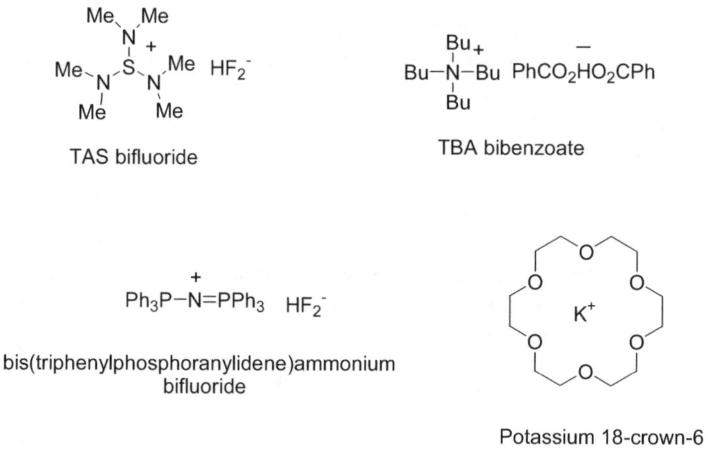

TAS bifluoride

TBA bibenzoate

bis(triphenylphosphoranylidene)ammonium bifluoride

Potassium 18-crown-6

Scheme 10

The catalysts are used at about 1–0.1% vs initiator concentration; in fact if too much catalyst is used the polymerization ceases [1].

3.4.2 *Lewis Acids*

Lewis acid catalysts were discussed in the acrylic monomer section. They are used at about 10% concentration vs monomer. Although no significant research has been done with them, other than looking for new ones, they most likely work by activating the monomers. Type Y zeolites catalyze GTP at about 25% concentration vs initiator. Conversions are quantitative and

MWDs as low as 1.02 were obtained [22]. Helmchen and Preston [23] report that P_4O_{10} in acetonitrile catalyzes the addition of silyl ketene acetals to enones in a high yield.

3.5 The GTP Mechanism

3.5.1 *GTP Phenomena that Must Be Accounted for by a Reasonable Mechanism*

- Ester enolates operate as both catalysts and initiators for GTP
- The need for low catalyst concentration
- A living methacrylate polymerization process that operates at 80 °C
- The reaction of catalyst with initiator
- The need for large unreactive counterions
- Induction periods
- 'Livingness' enhancing agents
- Rapid end group exchange in the presence of anionic catalysts
- Lack of exchange in double label experiments
- Reactivity ratios that differ from those of anionic and radical polymerizations
- Chain stereochemistry
- Monomer effects

The two main mechanistic routes for base catalyzed GTP under consideration are the dissociative pathway (Scheme 11) and the associative pathway (Scheme 12). In the dissociative route the nucleophilic catalyst complexes with the silyl ketene acetal end groups and in a reversible cleavage step gen-

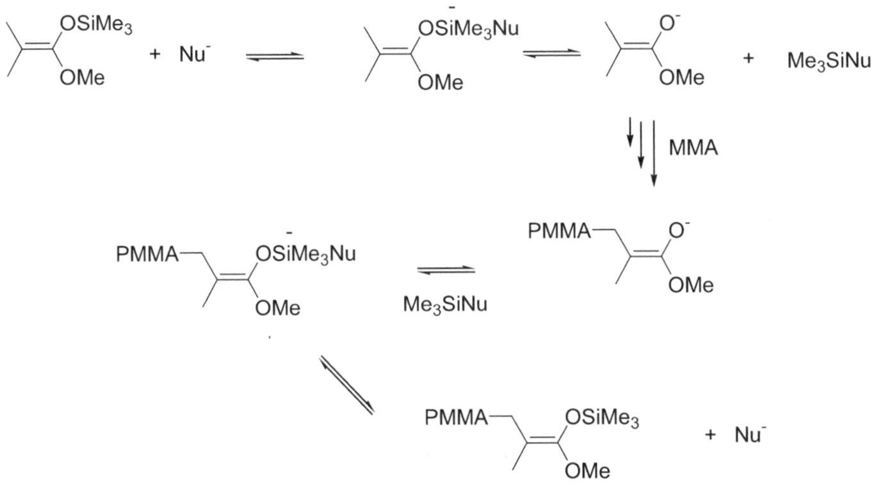

Scheme 11

erates a reactive enolate end that adds monomer. The enolate end groups are then capped by R_3SiNu to regenerate silyl ketene acetal ends. Since low MWD and controlled molecular weight polymer is obtained at low catalyst concentrations, the equilibrium generating enolate ends must be much faster than the rate of polymerization.

OMe
OSiMe3
+ Nu-
OMe
OSiMe3Nu
CO2Me
OSiMe3Nu
OMe
O
OMe
MMA
PMMA
OSiMe3
OMe
+ Nu-
PMMA
OMe
OSiMe3Nu

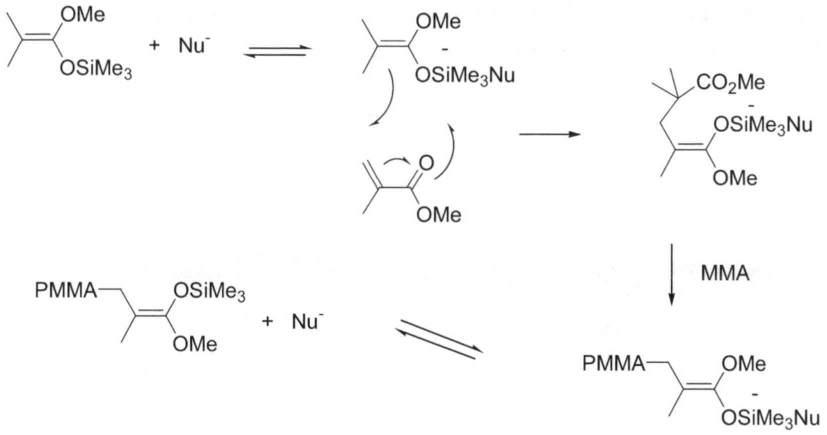

Scheme 12

In the associative mechanism the silyl ketene acetal group is activated by complexation with catalyst for addition to monomer. The silyl group transfers to incoming monomer and remains on the same polymer chain during the polymerization step. The well-documented silyl end group exchange would be occurring by some unknown process. The equilibrium rate for catalyst complex formation must be fast to insure molecular weight control and low MWD.

3.5.1.1
Ester Enolates Operate as Both Initiators and Catalysts for GTP

In a study that in itself almost lays to rest the associative mechanism for GTP, Quirk generated an ester enolate by addition of TBA 9-methylfluorenide to MMA and used it as a catalyst for GTP of MMA initiated by MTS at 50 °C. The conversion was quantitative, the MWD was in the 1.2 range, and the molecular weight was controlled by the ratio of MTS to monomer [6]. In similar experiments without the MTS, conversions of only 14–52% were obtained, the MWD was broad and molecular weight control was lost.

Quirk postulates that the ester enolate end groups are being stabilized by complexation with the silyl ketene acetal end groups (Scheme 13). He agrees that the complex could be adding monomer by the associative process but that since all chains are growing at the same rate, the equilibrium between complex and bare enolate must be faster than the rate of polymerization.

Thus unless the rate of polymerization of MMA by complex and by bare enolate where the same (which is highly unlikely) the MWD and Mn control would be lost.

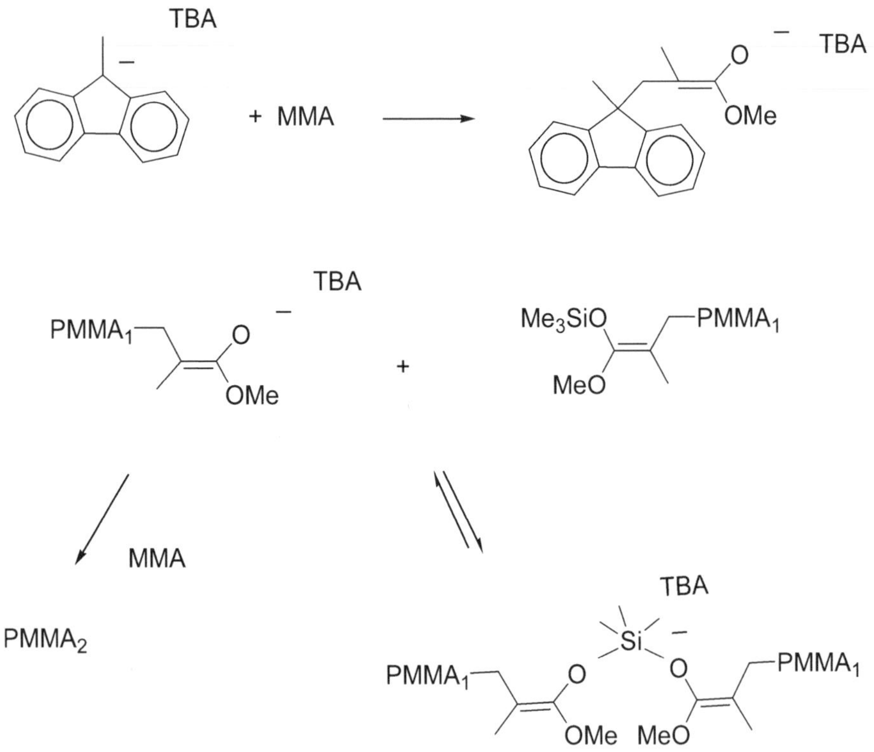

TBA

+ MMA

TBA

O

OMe

TBA

$PMMA_1$

O

OMe

+

Me_3SiO

$PMMA_1$

MeO

MMA

$PMMA_2$

TBA

Si

$PMMA_1$

O

O

$PMMA_1$

OMe MeO

Scheme 13

Quirk's argument is strengthened by Muller's finding that increasing the concentration of silyl ketene acetal retards GTP; the result of shifting the equilibrium between enolate and complex to the complex side and thus lowering the concentration of bare enolate and the rate of polymerization [24] (Scheme 13).

3.5.1.2
The Need for Low Catalyst Concentrations

The Quirk mechanism also provides us with an answer to a problem that has been with us since the inception of GTP: why does too much catalyst ruin the polymerization? In the associative mechanism increasing the amount of catalyst should merely increase the rate of reaction. However, in the dissociative mechanism increasing the amount of catalyst may generate more enolate than can be stabilized through complexation with the existing silyl ketene acetal end groups. (The equilibrium constant for association of one of the few pentavalent silicon complexes that has been studied, TBA $Me_3Si(CN)_2^-$, is 2.3 in THF [25]). Bandermann has studied the effect of catalyst concentration on GTP [26].

3.5.1.3
A Living Methacrylate Polymerization Process that Operates at 80 °C

Most of the early work on GTP was conducted by combining monomer, initiator and catalyst at room temperature and letting the reaction temperature rise to about 50 °C. Later process work was conducted in solvents at 80 °C so that the heat of polymerization could be removed by cooling with a reflux condenser. One of our reasons for proposing an associative GTP over a dissociative GTP was that an associative process would involve stable silyl ketene acetal groups as key intermediates in the formation of polymer. On the other hand the dissociative process would involve free ester enolate, known to be unstable at 80 °C. A closer look at the whole system however reveals why GTP works at above ambient temperatures, while classical anionic polymerization does not. *In GTP the backbiting termination reaction turns into a chain transfer process.* The backbiting reaction of chain ends produces a cyclohexanone derivative and methoxide. In classical anionic polymerization this terminates the chain. In GTP the methoxide can react with the latent silyl ketene acetal ends to regenerate enolate ends for further polymerization (Scheme 14).

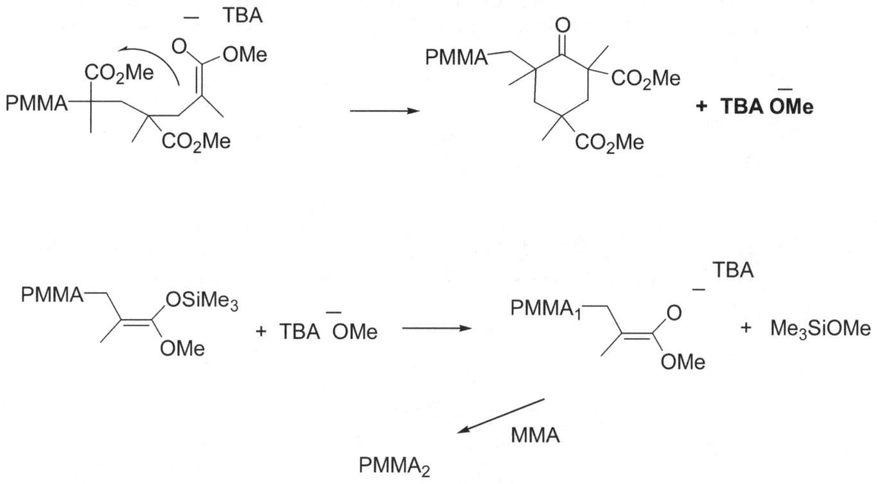

Scheme 14

For acrylate polymerization by GTP the cyclohexanone that results from backbiting is an α-ketoester with an active hydrogen that reacts with the methoxide thus thwarting the transfer process (Scheme 15).

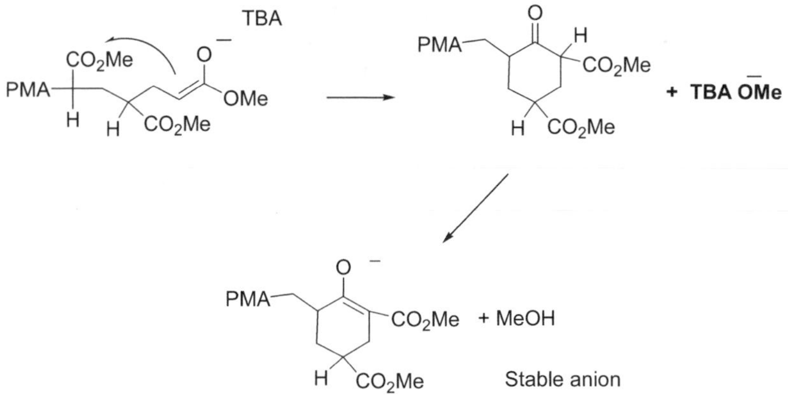

Scheme 15

3.5.1.4
The Reaction of the Catalyst with the Initiator

In the absence of monomer, GTP catalysts are destroyed by the initiator. A one to one molar mixture of bifluoride and MTS, for example produces methyl isobuterate, trimethylsilylfluoride, trimethylsilylmethoxide, methyl 2,2,4-trimethyl-3-oxopentanoate, and other oligomers [19, 26]. Seebach has shown that ester enolates generate ketenes at room temperature [27] (Scheme 16). These reactions support the dissociative mechanism wherein

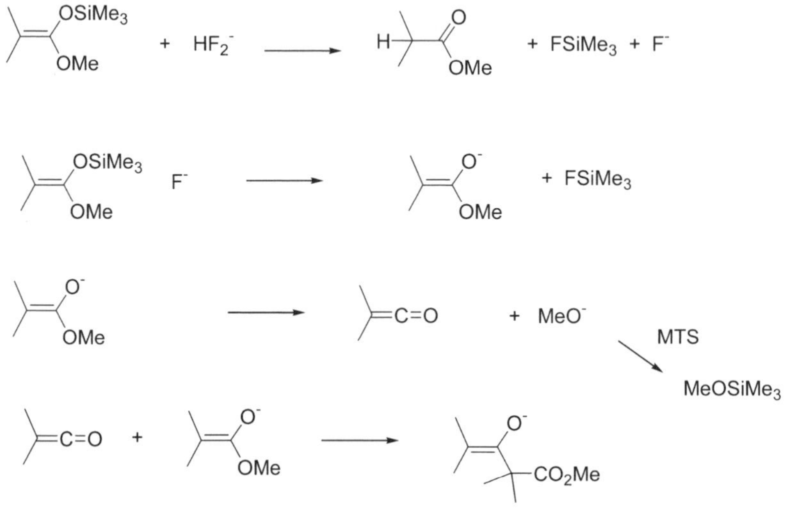

Scheme 16

ester enolate is generated. In the associative process the catalyst should merely complex reversibly with the MTS and not destroy it. At less than 1:1 ratios considerable isomerization of the ketene acetal to its C-silyl isomer occurs [19, 26]. At low concentrations of catalyst compared to initiator, the small amount of ester enolate generated would be stabilized by the presence of large amounts of MTS, therefore the catalyst and initiator can be combined before the addition of monomer.

3.5.1.5
The Need for Large Unreactive Counterions for Anionic GTP

In the initial studies on GTP TAS difluorosiliconate and bifluoride were found to be much better catalyst than alkali metal bifluorides. Later the more readily available TBA salts were used (Scheme 10). We assumed these large ion catalysts worked better because they formed stronger complexes with the silicon ketene acetal end groups by the associative mechanism. However, if one assumes a dissociative mechanism, then the large ions may work better by slowing down the rate of backbiting termination and formation of ketenes [27] (Schemes 14 and 16) from ester enolate ends. Reetz speculates that elimination of TBA alkoxide would be a higher energy process compared to elimination of metal alkoxide in these side reactions [13]. In addition in the dissociative process the large counterions would foster complex formation of enolate polymer ends with silyl ketene acetal ends.

3.5.1.6
Induction Periods

In many GTP experiments an induction period of up to a half an hour is observed before polymerization begins. This implies that during the induction period some inhibitor is being consumed or that some intermediate is formed that is the true initiator. In an associative process straightforward formation of catalyst/silylketene acetal complex should not require an induction period. On the other hand, in a dissociative process the catalyst first has to make a small amount of enolate.

An example of an inhibitor being consumed is seen in the initiation of GTP with TMS cyanide catalyzed by TBA CN^-. Here the polymerization does not begin until all of the TMS cyanide has added to the MMA (Scheme 17). Hertler studied the system by NMR and showed that Me_3SiCN complexes with the cyanide catalyst to the extent that polymerization will not take place [28] until the complex is gone. This complexation will certainly take place; however, the cyanide catalyst should continue to add to monomer until it is also gone leaving the polymer enolate complex (Scheme 17a), which then initiates polymerization.

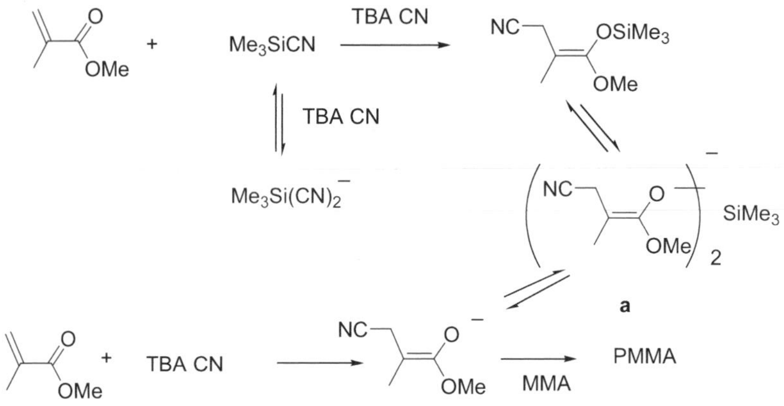

Scheme 17

3.5.1.7
'Livingness' Enhancing Agents

Early in our studies on GTP we discovered that certain additives lowered MWDs and gave better molecular weight control (livingness enhancement). They also lowered the rate of polymerization. Some examples are: $ArCO_2$-$SiMe_3$, $MeCO_2SiMe_3$, MeCN, and Me_3SiNMe_2. In the associative mechanism these agents would be complexing with catalyst to lower its concentration and thereby lower the rate of polymerization (Scheme 18a). In the dissocia-

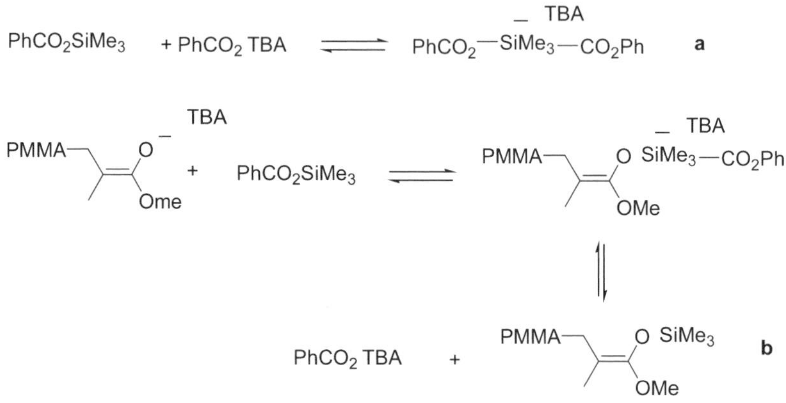

Scheme 18

tive mechanism they would be complexing with enolate chain ends as well as with the catalyst. The trimethylsilyl carboxylates [29] are a special case since these reagents also possess the ability to silylate the reactive enolate ends (Scheme 18b). Brittain showed that at a concentration of trimethylsilyl benzoate equal to that of catalyst the rate of polymerization is lowered 50-fold [30] (the TMS benzoate was produced by the in situ reaction of TBA bibenzoate with MTS). At a concentration equal to five times that of initiator, the rate is lowered still further [31] (Scheme 18). Similar results were obtained with TMS Ac [32].

3.5.1.8
Rapid End Group Exchange in the Presence of Anionic Catalysts

Early work on the GTP mechanism showed that the silyl groups on chain ends rapidly exchange in the presence of anionic catalysts [33, 34]. Without catalyst no exchange occurs [35]. No exchange occurred in the bifluoride catalyzed polymerization of MMA with dimethylphenylsilyl ketene acetal (Scheme 19a) in the presence of dimethyltolylsilyl fluoride [1]. However, in a similar experiment with trimethylsilyl acetate, TBA Ac, and dimethylphenylsilyl ketene acetal, complete exchange occurred within 5 min [36] (Scheme 19b).

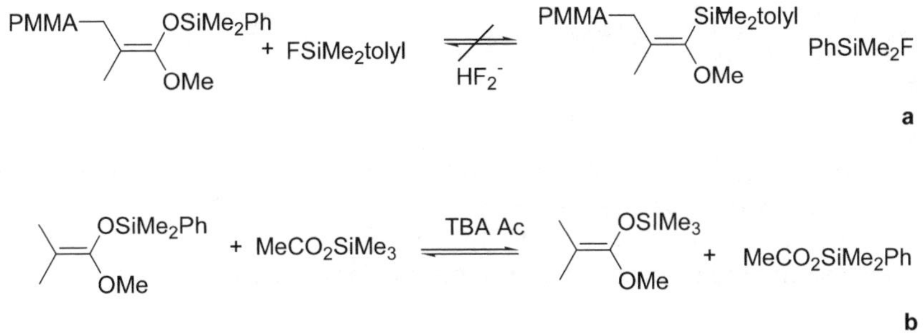

PMMA— $C(Me)=C(OSiMe_2Ph)(OMe)$ + $FSiMe_2$tolyl —(HF_2^-, no exchange)→ PMMA— $C(Me)=C(SiMe_2tolyl)(OMe)$ + $PhSiMe_2F$ **a**

$Me_2C=C(OSiMe_2Ph)(OMe)$ + $MeCO_2SiMe_3$ ⇌(TBA Ac) $Me_2C=C(OSiMe_3)(OMe)$ + $MeCO_2SiMe_2Ph$ **b**

Scheme 19

In the associative mechanism this exchange would be a side reaction not related to the polymerization process. The exchange would have to occur by a four centered transition state between two chain ends at least one of which is complexed to catalyst.

In the dissociative mechanism the exchange is readily explained by the formation and dissociation of the enolate ends with neutral silyl ketene acetal ends (Scheme 13). The lack of exchange of fluorosilane with enolate ends could be caused by the complex with fluorosilane breaking only at the SiO bond to revert to fluorosilane (no exchange).

3.5.1.9
The Lack of Exchange in Double-Label Experiments

The main evidence for the associative mechanism consists in double-labeling experiments conducted by Farnham and Sogah [33, 34]. At −90 °C in THF a mixture of PMMA with a dimethyltolylsilyl ketene acetal end group and PBMA with a dimethylphenylsilyl ketene acetal end group was treated with a small amount of BMA and TASF catalyst (Scheme 20). After 5 min the polymerization was quenched and the polymers separated by solubility. NMR showed a small PBMA block on the PMMA polymer but the end group was only tolyldimethylsilyl.

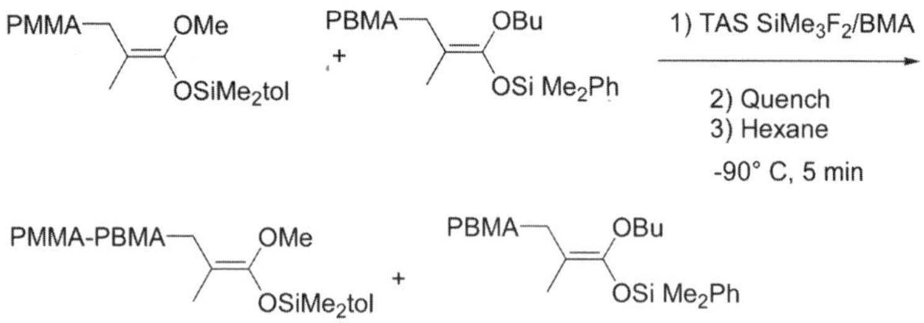

PMMA–C(CH₃)=C(OMe)(OSiMe₂tol) + PBMA–C(CH₃)=C(OBu)(OSi Me₂Ph) → [1) TAS SiMe₃F₂/BMA; 2) Quench; 3) Hexane; -90° C, 5 min] → PMMA-PBMA–C(CH₃)=C(OMe)(OSiMe₂tol) + PBMA–C(CH₃)=C(OBu)(OSi Me₂Ph)

Scheme 20

At −70 °C complete scrambling of end groups occurred. It thus appears that at −90 °C an associative mechanism is operating but at −70 °C a dissociative mechanism takes over (Scheme 21). In any case, polymerization at −90 °C is impractical and the room temperature mechanism is the important one.

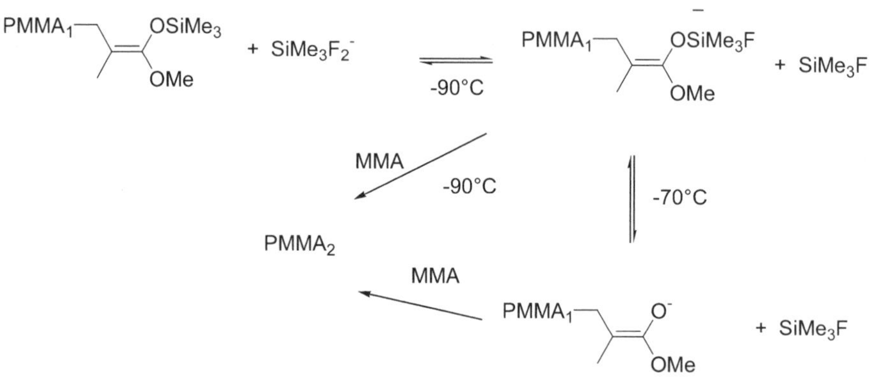

$PMMA_1$–C(CH₃)=C(OSiMe₃)(OMe) + $SiMe_3F_2^-$ ⇌ (-90°C) [$PMMA_1$–C(CH₃)=C(OSiMe₃F)(OMe)]$^-$ + $SiMe_3F$

[$PMMA_1$–C(CH₃)=C(OSiMe₃F)(OMe)]$^-$ → (MMA, -90°C) → $PMMA_2$

[$PMMA_1$–C(CH₃)=C(OSiMe₃F)(OMe)]$^-$ ⇌ (-70°C) $PMMA_1$–C(CH₃)=C(O$^-$)(OMe) + $SiMe_3F$

$PMMA_1$–C(CH₃)=C(O$^-$)(OMe) → (MMA) → $PMMA_2$

Scheme 21

Similar results were obtained with PBMA and PMMA tagged with triethylsilyl and trimethylsilyl groups. However here Bywater has noted that the triethylsilyl ended polymer may not be polymerizing significantly during the 5 min before quenching. Thus its silyl group would remain with its chain [19].

In double-labeling experiments using PMMA of two different molecular weights Quirk obtained partial exchange in 2 h [37]. Since the dissociate mechanism demands complete exchange, these experiments support an associative pathway.

Evidence favoring an associative mechanism was obtained in dual initiator studies. Under an associative mechanism with two different initiators in the same reactor each set of chains would grow at slightly different rates. Thus the MWD of the resulting polymer should be higher than the one with one initiator. This is the case when dimethylphenylsilyl ketene acetal and TMS were used to polymerize MMA with TBA biacetate as catalyst [38] (Scheme 22).

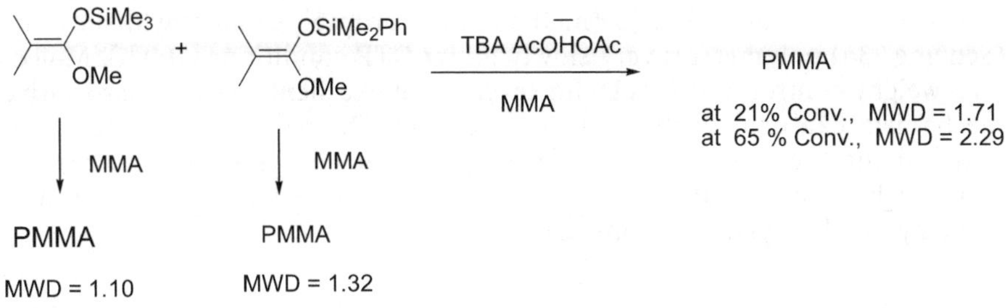

OSiMe$_3$ / OMe + OSiMe$_2$Ph / OMe —(TBA AcOHOAc$^-$, MMA)→ PMMA

at 21% Conv., MWD = 1.71
at 65 % Conv., MWD = 2.29

OSiMe$_3$ / OMe —(MMA)→ PMMA, MWD = 1.10

OSiMe$_2$Ph / OMe —(MMA)→ PMMA, MWD = 1.32

Scheme 22

3.5.1.10
Reactivity Ratios that Differ from those of Anionic and Radical Polymerizations

Haddleton determined the reactivity ratios for copolymerization of MMA with BMA by classical anionic as 1.04: 0.81: by alkyllithium/trialkylaluminum initiation, 1.10: 0.72; by GTP, 1.76: 0.67: by ATRP, 0.98: 1.26; by catalytic chain transfer, 0.75: 0.98; by classical free radical, 0.93: 1.22 [39]. The difference in reactivity ratios between GTP and classical anionic polymerization seems to indicate GTP is an associative process. However, Jenkins has also measured reactivity ratios for the same pair by GTP and reports different results: r_{MMA}=0.44 and r_{BMA}=0.26 [40].

3.5.1.11
Chain Stereochemistry

The stereochemistry of GTP of MMA polymerization was measured for Lewis acid as well as for bifluoride catalysis. Lewis acid catalysts gave a ratio of syndiotactic:heterotactic triads of 2:1 while bifluoride catalysis gave ratios near 1:1 [9, 41]. The amount of isotactic triads was about 5%. The effect of temperature on triad and diad composition provided data to calculate the difference in activation enthalpy ($\Delta\Delta H^{\#}$) and activation entropy ($\Delta\Delta S^{\#}$) for m and r diad formation. The ($\Delta\Delta H^{\#}$) favoring r formation is only 1.0 kcal/mol while the ($\Delta\Delta G^{\#}$) favoring m formation is 1.1 eu.. At 273 °C: G=0.70 kcal/mol. Nearly the same values were found for free radical polymerization [42]. The amount of isotactic triads for GTP of *tert*-butyl methacrylate rose to 21% [43].

The differences between Lewis acid and anionic catalyzed GTP are expected. The similarities between radical initiated and GTP of MMA indicates chain stereochemistry will not contribute any answers to the mechanistic problem.

3.5.1.12 Monomer Effects

Certain monomers provide evidence relating to the GTP mechanism problem. Ober has shown that [3-(methacryloxy)propyl]pentamethyldisiloxane, (Scheme 23a) polymerizes very slowly under GTP conditions (100 h). Molecular weight control and MWD, however, were excellent [44]. A dissociative mechanism can explain this unusual property. The tethered siloxane group is stabilizing the enolate ends by forming a cyclic complex (Scheme 23). No mention has been made of a similar stabilizing effect for 2-trimethylsilyloxyethyl methacrylate (Scheme 23b).

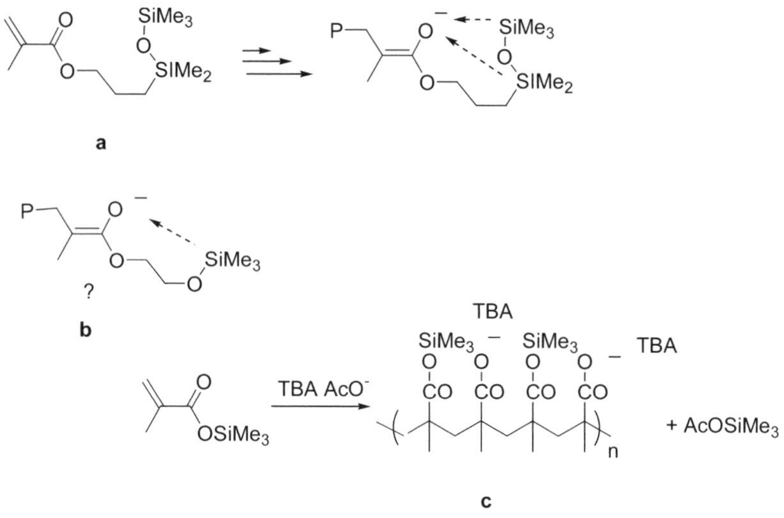

Scheme 23

The fact that trimethylsilyl methacrylate is a sluggish monomer under GTP conditions [45, 46] also bodes well for a dissociative mechanism. The excess silyl carboxy groups are silylating enolate chain ends Thus lowering the rate of polymerization and changing the nature of the carboxylate catalyst (Scheme 23c).

3.5.1.13
Kinetic Studies Relating to GTP

GTP kinetics have been studied by Brittain [30] using stopped flow FT-IR, by Muller [24, 47, 48] using gravimetrics, and by Bandermann [49–51] using dilatometry. Based on what we now know about GTP these studies are clouded by several factors. The catalysts are very likely reacting irreversibly with the initiators to produce products that are involved in the polymerization: TAS fluoride makes ester enolate plus silyl fluoride (Scheme 16). Bifluoride makes ester enolate plus silyl fluoride plus methyl isobuterate (Scheme 16). TBA bibenzoate makes silyl benzoate plus TBA benzoate [31]. (Therefore kinetic studies on bibenzoate are really studies on benzoate plus one equivalent of silyl benzoate). These are known "side" reactions. What is critical, however, is how fast do the "side" reactions occur? Do they happen during the induction periods? Are they over before significant polymerization takes place?

One thing everyone more or less agrees on is that GTP is first order in monomer and in catalyst. A significant result by Muller is that the initiator has a slight retarding effect on GTP [24]. Brittain has found that the rate of polymerization to be 50 times slower for bibenzoate than for benzoate catalysis [30]. This means that silyl benzoate has a retarding effect (most likely due to both complexation with catalyst and re-silylation of ester enolate end groups) (Scheme 18). In a similar manner, bifluoride may be lowering the rate of polymerization by providing silyl fluoride that complexes with enolate chain ends.

The relative order for catalyst activity is $F^- > HF_2^- > Ac^- > Bz^- > Bz_2H^-$ [30].

3.5.2
Conclusions Relating to the Mechanism of GTP

The fact that known anionic initiators for MMA can act as catalysts for GTP and the need for low amounts of catalysts in itself nearly puts to rest the associative mechanism. Seven of the other factors support the dissociative process. Except for the low temperature exchange studies, none supports the associative mechanism. Based on the lack of exchange of added silyl fluoride with silyl ketene acetal ends it looks like fluoride and bifluoride catalysts operate by *irreversible* generation of ester enolate chain ends [1] (Scheme 19b). On the other hand carboxylate catalysts appear to operate by *reversible* generation of ester enolate ends as evidenced by rapid exchange of silyl acetate with silyl ketene acetal ends [36] (Scheme 19c).

4
Commercial Uses for GTP

GTP is currently being used by DuPont to manufacture dispersing agents for pigments and for emulsion stabilizers. Water-based dispersants are used in jet printer inks and organic-solvent-based dispersants for auto finishes. Less pigment is needed to reach the desired color intensity for auto finishes thus offsetting the cost of the dispersing agent. Spinelli has published limited details relating to the water based dispersants [52]. The AB block block polymers have a balance of hydrophobic 'A' blocks and very hydrophilic 'B' blocks that are composed of neutralized acid or amine containing methacrylates.

5
Comparison with Other Systems

In the early years of the polymer age, 1920–1950, free radical vinyl polymerization was used to manufacture homo- and random copolymers of ethylene, propylene, styrene, vinyl chloride, butadiene, and members of the acrylic family of monomers. In the mid-1950s Szwarc [53] opened a new era in polymerization technology with his recognition of the living nature of the anionic polymerization of styrene and isoprene. Similar techniques were soon found for acrylic monomers [7]. However the flood gates for controlled polymerization were really opened in the 1980s by Higashimura and Sawmoto's discovery of living cationic polymerization of vinyl ethers [54], by Kennedy's living cationic polymerization of isobutylene [55], and by the DuPont group's discovery of GTP for polymerization of acylic monomers [1].

Since GTP deals mainly with the acrylic family of monomers, comparison to other controlled methods for polymerization of this class will be covered. The huge, as yet unrealized, commercial potential of controlled free radical polymerization is of special note.

5.1
Immortal Polymerization

Inoue [56] has developed a method similar to GTP for polymerization of acrylic monomers. A methylaluminum porphyrin (MeAlTPP) is converted to a ketene acetal by in situ reaction of MMA and used to polymerize MMA (Scheme 24). A hindered Lewis acid catalyst is needed to activate the MMA.

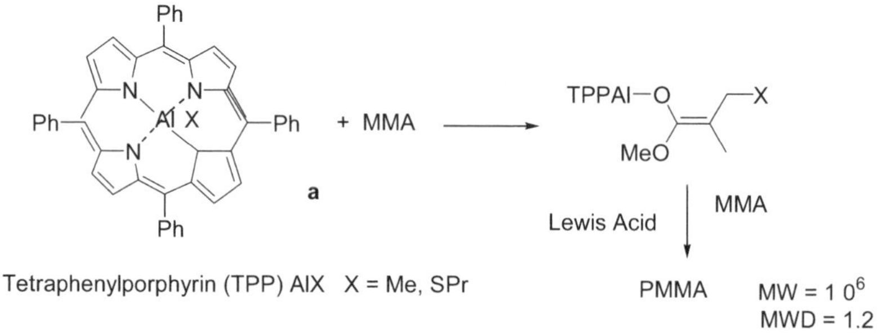

+ MMA → TPPAl—O(MeO)C=C(CH$_3$)CH$_2$—X

a

Lewis Acid, MMA ↓

PMMA MW = 1 0^6 MWD = 1.2

Tetraphenylporphyrin (TPP) AlX X = Me, SPr

Scheme 24

No separate step to form the enolate is required if an alkylthio TPP is used. High molecular weight polymer with low MWD is formed. The method was first used to polymerize oxiranes. No industrial applications have been reported, possibly due to the initiator expense and its deep red color. Immortal polymerization has been reviewed [57, 58].

5.2 Rare Earth Enolates as Initiators

Yasuda [59, 60] has found that $(Cp^*)_2SmH$ ($Cp^*=C_5Me_5$) will initiate living polymerization of MMA. Very high molecular weight polymer with low MWDs is produced (Scheme 25a). A block polymer with ethylene has been made. Novak [61] claims that a dimer intermediate first forms (Scheme 25b).

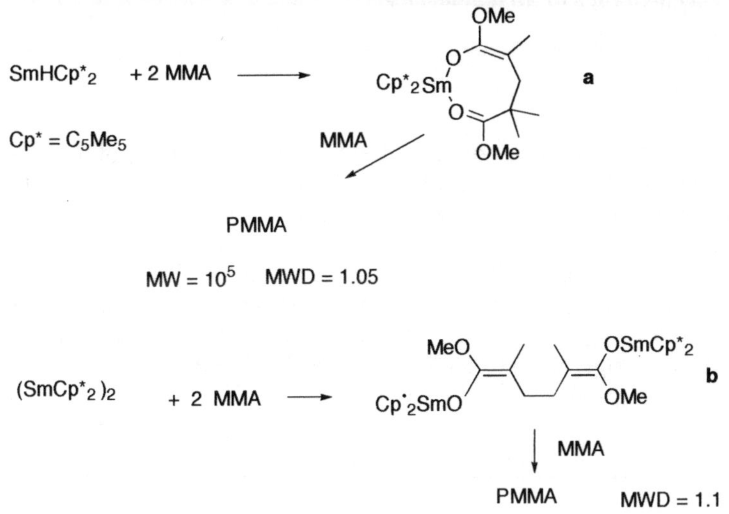

$SmHCp^*_2$ + 2 MMA → Cp^*_2Sm (cyclic bis-enolate, OMe, OMe) **a**

$Cp^* = C_5Me_5$

MMA → PMMA

MW = 10^5 MWD = 1.05

$(SmCp^*_2)_2$ + 2 MMA → Cp^*_2SmO(MeO)C=C…C=C(OMe)$OSmCp^*_2$ **b**

MMA → PMMA MWD = 1.1

Scheme 25

This method looks good but the extreme difficulty in handling the unstable initiators may be holding back its use other than for research. The cost of the C_5Me_5 ligand may also be a problem.

5.3 Other Transition Metal Initiators

Collins [62] has shown that $Cp_2ZrMe\ BPh_4$ will also initiate polymerization of MMA. PMMA with Mn in the 100,000 can be made (Scheme 26). Cost may be holding back development of these types of initiators.

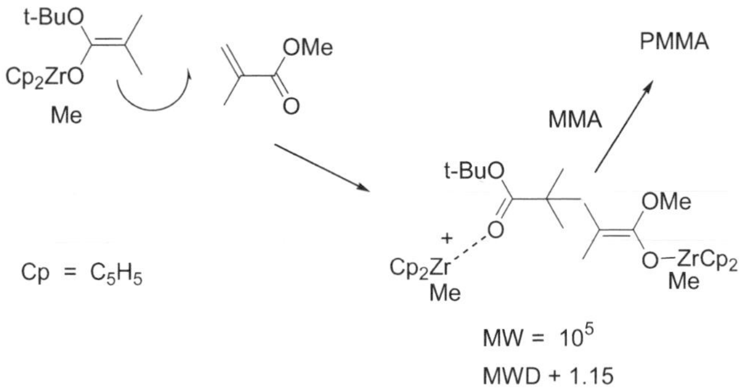

Scheme 26

5.4
Anionic Polymerization of Acrylic Monomers

5.4.1
Classical Living Anionic Polymerization

In the mid 1970s DuPont conducted research on anionic polymerization of methacrylates [7] to produce block polymer dispersing agents. Diphenylhexyl lithium was used as the initiator at below −60 °C (Scheme 27). At these temperatures the bulky initiator does not react with the ester groups on the monomers and the backbiting reaction (Scheme 3) is frozen out. Molecular weight control and low MWDs were obtained and the products were excellent dispersing agents, but the cost of purifying solvents and monomers and the need for refrigeration to cool reactors killed the project.

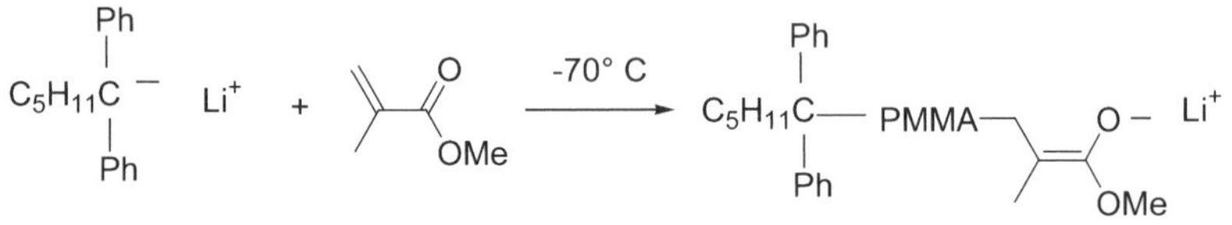

Scheme 27

5.4.2
Ligated Living Anionic Polymerization

Teyssie and Jerome [63] have solved the temperature problem to some extent by conducting anionic polymerization of acrylates and methacrylates in the presence of LiCl, LiOR, or $LiO(CH_2CH_2O)_nMe$. These agents complex with lithium enolate chain ends.

Ballard et al. [64] found that bulky dialkyl aluminum phenolate additives would improve the anionic polymerization of acrylic monomers. They called their method Screened Anionic Polymerization (Scheme 28).

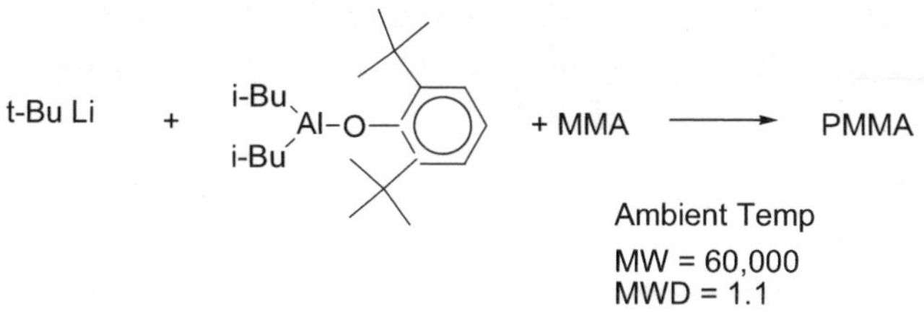

t-Bu Li + i-Bu, i-Bu Al–O–(2,6-di-t-butylphenyl) + MMA ⟶ PMMA

Ambient Temp
MW = 60,000
MWD = 1.1

Scheme 28

5.4.3
Tetraalkylammonium and Other Bulky Counterions for Anionic Polymerization

Reetz [13, 65] and also Sivaram [66] have shown that nucleophilic tetrabutylammonium salts will initiate living polymerization of acrylates at room temperature. Molecular weights 1500–25,000 are obtained with MWDs of 1.1–1.4. Methyl and ethyl acrylates don't work as well as the more bulky acrylates. Side reactions are end group cyclization and Hoffmann elimination [13] (Scheme 29).

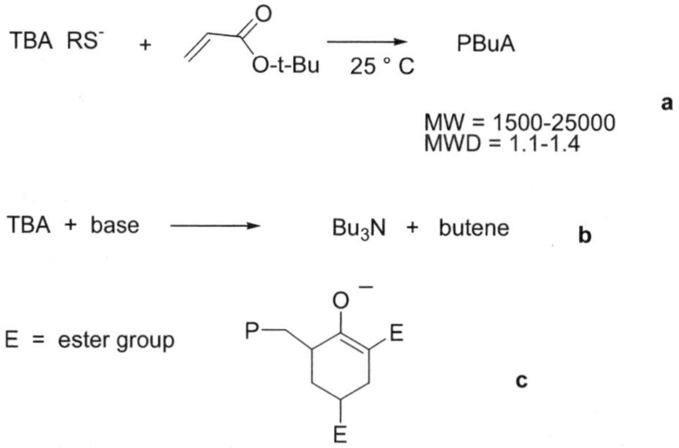

TBA RS^- + CH_2=CH–C(=O)O-t-Bu ⟶ (25 ° C) PBuA **a**

MW = 1500-25000
MWD = 1.1-1.4

TBA + base ⟶ Bu_3N + butene **b**

E = ester group **c**

Scheme 29

We have found that the size of the gegenion is more important than the fact that it is nonmetallic. Potassium dimethyl malonate/18-crown-6 polymerizes MMA at 25–60 °C to give quantitative yields of PMMA, MWD 1.5–1.9. Excess malonate or methanol lowered the molecular weight of the PMMA but did not shut down the polymerization [67]. It is tempting to postulate that a hydrogen-bonded version of the Quirk intermediate (Scheme 13) is stabilizing enolate ends (Scheme 30).

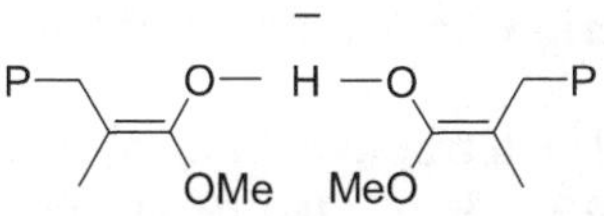

P– –O– H –O– –P
OMe MeO

Scheme 30

Haggard and Lewis have shown that tetrabutylammonium alkoxides polymerize methacrylates in alcohol solution [68]. Block polymer synthesis is not possible since the ester groups on the monomer exchange with the solvent.

Seebach [69] used P_4 base (Scheme 31a) as an initiator for the anionic polymerization of MMA. Good control of molecular weight and MWDs are realized at temperatures up to 60 °C (Scheme 31a). The process has the strange property of not proceeding at temperatures below −20 °C (all other anionic polymerizations of MMA work better at low temperatures). The experimental MWs are higher than those expected by the amount of P_4 base used. A. Muller has confirmed these results using P_5 counterion at 20 °C in a tubular reactor [70] (Scheme 31b).

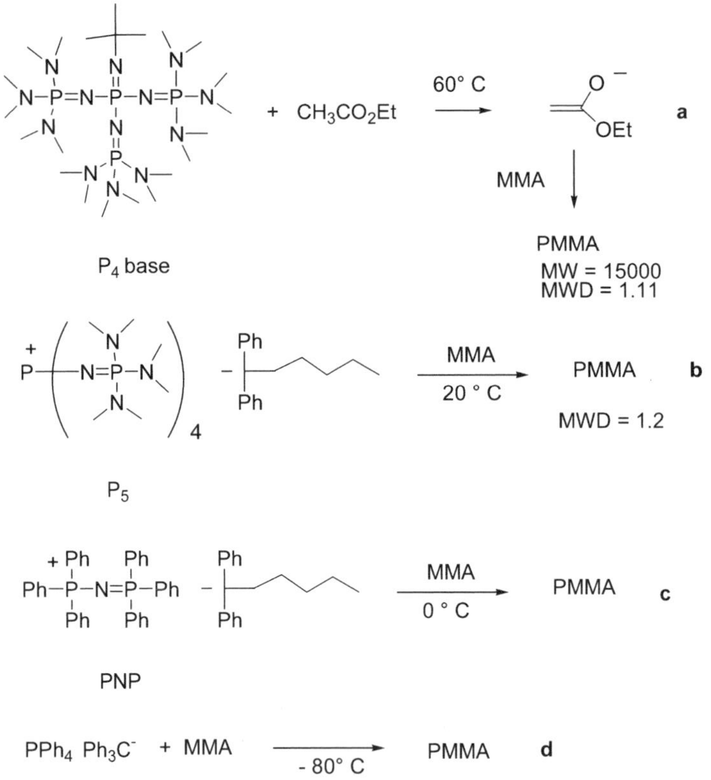

P_4 base + CH_3CO_2Et —(60° C)→ enolate (O⁻, OEt) —(MMA)→ PMMA, MW = 15000, MWD = 1.11 **a**

P_5 + Ph₂C⁻ (alkyl) —(MMA, 20 °C)→ PMMA, MWD = 1.2 **b**

PNP + Ph₂C⁻ (alkyl) —(MMA, 0 °C)→ PMMA **c**

$PPh_4\ Ph_3C^-$ + MMA —(- 80° C)→ PMMA **d**

Scheme 31

H. Muller has used the related bis(triphenylposphoranylidene)ammonium ion (PNP) as a counterion for polymerization of MMA at 0 °C [71] (Scheme 31c).

Hogen-Esch [72] used a tetraphenylphosphonium gegenion for anionic polymerization of MMA. Low temperatures (−80 °C) were used

(Scheme 31d), however, A. Muller has shown that –20 to 20 °C is adequate [73].

Anionic polymerization of (meth)acrylates with hindered ester functions can most likely be conducted at room temperature and above to remove the heat of polymerization with low boiling solvents. Polymerization of the important methyl and ethyl (meth)acrylate members of the family, however, are still plagued by chain termination at higher temperatures. The phosphorus based counterions have a stability advantage over tetraalkylammonium counterions which undergo Hoffman elimination.

5.5 Controlled Free Radical Polymerization of Acrylic Monomers

The goal of producing *low cost* ($1–5/lb.) acrylic block, comb, star, and telechelic polymers by GTP and anionic polymerization has not been met. Free radical polymerization of acrylics and other vinyl monomers on the other hand requires little purification of materials, works in water and other protic solvents, and is low cost. Considerable efforts are presently under way therefore to develop controlled free radical polymerization methods.

Early work by Otsu [74] and by Braun [75] showed some promise using Me_2NCS_2 and triphenylmethyl radicals as reversible capping agents for free radical chain ends. But the capping agents themselves initiated monomer and were thus slowly depleted. The iniferter concept has been reviewed by Otsu [76].

5.5.1 *Nitroxide as a Reversible Cap for Free Radical Polymerization*

The first workable capping agents for controlled radical polymerization were discovered by Rizzardo et al. [77, 78] who used nitroxides. The nitroxide reacts reversibly with radical chain ends but itself does not initiate the monomer. They called their new system Stable Free Radical Polymerization (SFRP). Scheme 32a depicts an example of SFRP using TEMPO (2,2,6,6-tetramethyl-1-piperidinyloxy). SFRP was developed independently by Georges at Xerox for the synthesis of styrene block polymer as dispersing agents [79].

While SFRP has the advantage of being all organic (no metals or halogen to cause corrosion problems), it is too slow for an ideal industrial process. On average to get suitable conversions, polymerizations take about 24 h. Camphor sulfonic or benzoic acids have been use as catalysts to increase the rate of polymerization [79]. The system operates at 100–120 °C with good control of polymer molecular weight. MWDs in the 1.1 range are obtained. Block polymers are possible but excess monomer must be removed before proceeding with the formation of the second block.

Hawker has done a combinatorial study to find the best nitroxide [80]. His work shows that nitroxide (Scheme 32b) is the best SFRP capping agent available at this time. He proposes that it is better since it is not as stable as

TEMPO. Thus it is depleted from the polymerization solution at about the same rate that chain ends are being depleted by combination. Without this depletion the nitroxide concentration builds up and lowers the polymerization rate. A similar "unstable" nitroxide has been reported by Gnanou [81] (Scheme 32c). Hawker has reviewed SFRP [82].

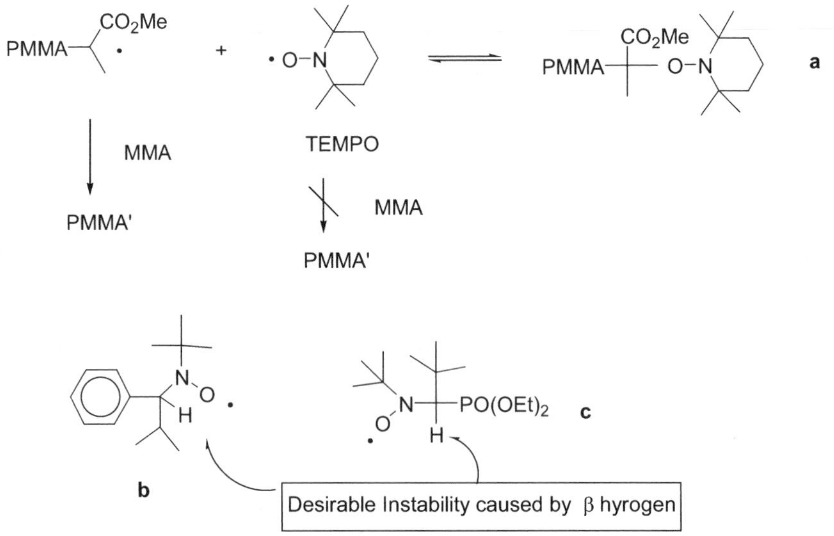

Scheme 32

5.5.2
Atom Transfer Polymerization (ATRP)

Matyjaszewski [83] and Sawamoto [84] have found that α-halo ester chain ends can be used to generate radical ends reversibly by treatment with copper complexes or $RuCl_2/MeAl(ODBP)_2$ (ODBP=ortho-di-*tert*-butylphenoxy) (Scheme 33a,b). In his early work Sawamoto used carbon tetrachloride as initiator and Matyjaszewski, α-halo esters. Percec [85, 86] discovered that sulfonyl chlorides provided advantages over the other initiators and has used these initiators extensively in his research (Scheme 33c). ATRP has been fine-tuned by the three groups. The reader is directed to reviews by Matyjaszewski [87] and by Sawamoto [88].

ATRP gives excellent control of polymer chain architecture. For industrial use, however, two problems need to be overcome: residual halides and metals in the product would be a problem for electronic device uses. The rate of polymerization may be too slow. This is because the chain end concentrations must be low so that typical radical chain termination is kept to a minimum. Chain termination is a second order reaction and will be minimized by low concentrations of chain end radicals. The low rate of polymerization may increase the cost of the process since the optimum time for a polymerization run is about 6 h.

$Cl{-}C(CH_3)_2{-}CO_2Me + CuCl \rightleftharpoons {\cdot}C(CH_3)_2CO_2Me + CuCl_2 \xrightarrow{MMA} PMMA$ **a**

$CCl_4 + RuCl_2 \xrightarrow[MeAl(ODB\ P)_2]{MMA} Cl_3C{-}C(CH_3)(CO_2Me){-}CH_2{-}Cl \rightleftharpoons {\cdot}C(CH_3)(CO_2Me){-} \quad RuCl_3$ **b**

ODRP = 2,6-di-t-butylphenoxy

$C_6H_5{-}SO_2Cl + CuCl \xrightleftharpoons{MMA} C_6H_5{-}SO_2{-}C(CH_3)(CO_2Me){-}Cl \xrightarrow{MMA} PMMA$ **c**

Scheme 33

5.5.3 *Reversible Addition Fragmentation Transfer (RAFT)*

Rizzardo et al. [89, 90] has developed a method for the controlled polymerization of acrylic monomers that involves the addition of a radical chain end to other chain ends capped with dithioester functions. After addition the resulting stabilized radical fragments back to the same radical that added or to a new fragment. Either fragment can then add monomer. In essence, the process is a reversible transfer of polymer chain ends. They named it Radical Addition Fragmentation Transfer (RAFT). At any one time all but a trace of the end groups are dithioesters (Scheme 34). Although excellent molecular weight and MWDs are obtained the product is smelly and yellow colored. Davis [91] claims the odors disappear on boiling in THF containing a little peroxide.

$R{-}C(=S){-}S{-}C(CH_3)(CO_2Me){-}PMMA' + PMMA{-}C(CH_3)(CO_2Me){\cdot} \rightleftharpoons PMMA{-}C(CH_3)(CO_2Me){-}S{-}\dot{C}(R){-}S{-}C(CH_3)(CO_2Me){-}PMMA'$

$\rightleftharpoons PMMA{-}C(CH_3)(CO_2Me){-}S{-}C(=S){-}R + {\cdot}C(CH_3)(CO_2Me){-}PMMA'$

${\cdot}C(CH_3)(CO_2Me){-}PMMA' \xrightarrow{MMA} PMMA''$

Scheme 34

5.5.4
1,1-Diphenylethylene as a Reversible Cap

In a process related to RAFT, BASF workers have shown that 1,1-diphenylethylene will control the molecular weight of PMMA and polystyrene, and permit block polymer synthesis [92]. They propose that radical chain ends add to the diphenylethylene to form a stable diphenylalkyl radical that does not add more monomer but can reverse to diphenylethylene and the same radical chain end for addition of more monomer. The diphenylalkyl radical cap has the additional possibility of forming a reversible dimer (Scheme 35).

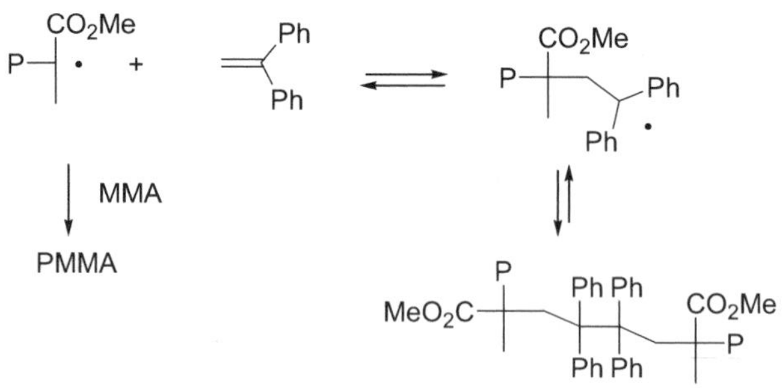

CO_2Me P • + Ph Ph ⇌ CO_2Me P Ph Ph •

MMA

PMMA

MeO_2C P Ph Ph Ph Ph CO_2Me P

Scheme 35

Details of the new process are minimal but it looks promising for commercial use.

5.5.5
Catalytic Chain Transfer (CCT)

Although it is not a living polymerization CCT has the ability to control the molecular weight of PMMA and to end up with a vinyl end group. These macromonomers are suitable for copolymerization with other monomers to form comb polymers. The technique [93.94] uses a cobalt tetraphenylporphrin or glyoxime to remove a hydrogen atom from the methyl group of radical chain ends and transfer it to monomer to start new chains (Scheme 36a). CCT is presently being used commercially to manufacture medium molecular weight PMMA.

Acrylates (which have no methyl group adjacent to the radical ends during polymerization) are reversibly capped by cobalt tetraphenylporphrin resulting in a living polymerization [95] (Scheme 36b).

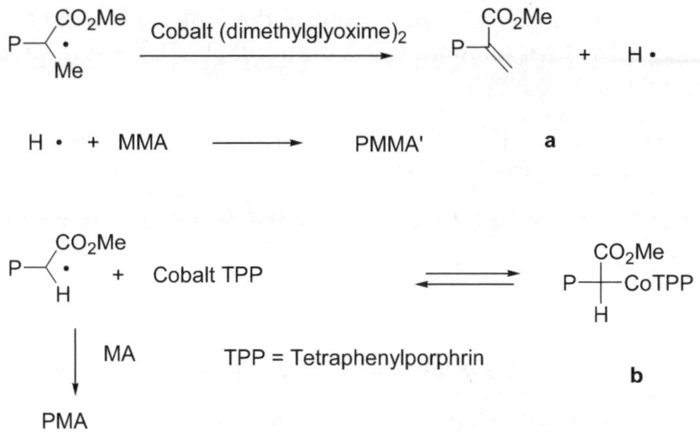

P–C•(CO_2Me)(Me) $\xrightarrow{\text{Cobalt (dimethylglyoxime)}_2}$ P–C(CO_2Me)=CH_2 + H•

H • + MMA ⟶ PMMA' **a**

P–C•(CO_2Me)(H) + Cobalt TPP ⇌ P–C(CO_2Me)(H)–CoTPP **b**

↓ MA

PMA

TPP = Tetraphenylporphrin

Scheme 36

6
Conclusions

With respect to living polymerization of meth(acrylates), GTP is the only technique being used commercially. Block polymer dispersing agents for pigmented water based inks and for dispersing pigment in polymer resins are the products. These hi-tech uses can afford the additional cost to run GTP vs traditional polymerization procedures. For other uses lower cost methods are still needed.

Classic anionic polymerization must be conducted at too low temperature for commercial feasibility. Anionic polymerization of hindered acrylates and methacrylates with stable large counterions works at temperatures close to those required for condenser cooling. The anionic polymerization of lower alkyl acrylates and methacrylates is still a problem.

SFRP takes too much time to complete the polymerization run and the nitroxide end reagents are expensive and unstable.

The products from ATRP will most likely contain metallic and halide impurities. The sulfonyl halide initiators for ATRP have considerable advantages over alkyl halides in cost and in operating with any monomer capable of undergoing ATRP.

RAFT produces polymers that contain foul smelling by-products that must be removed by a separate oxidation step.

In all the free radical procedures, to obtain complete monomer consumption before the addition of the second monomer to make a pure second block is a problem.

Reversible addition of 'unpolymerizable' monomers for end capping shows considerable promise.

Catalytic chain transfer is used to control the molecular weight of PMMA commercially. The macromonomers produced by CCT show promise for synthesis of comb polymers.

Acknowlegments The bulk of the DuPont work described herein was conducted by research workers in the Automotive Finishes and Central Research Departments. I wish to especially thank Dotsi Sogah, Wally Hertler, Bill Farnham, T.V. RajanBabu, and Professors Jack Roberts and Barry Trost for many hours of discussion on various aspects of GTP technology.

References

1. Webster OW, Hertler WR, Sogah DY, Farnham WB, RajanBabu TV (1983) J Am Chem Soc 105:5706
2. Eastmond GC, Webster OW (1991) Group transfer polymerization. In: Ebdon JR (ed) New methods of polymer synthesis. Blackie, London, p 24
3. Webster OW, Anderson BC (1992) Group transfer polymerization. In: Mijs WJ (ed) New methods for polymer synthesis. Plenum, New York, p 1
4. Hertler WR (1996) Group transfer polymerization. In: Kricheldorf HR (ed) Silicon in polymer synthesis. Springer, Berlin Heidelberg New York, p 69
5. Hertler WR (1997) Group transfer polymerization for controlled polymer architecture. In: Hatada K, Kitayama T, Vogl O (eds) Macromolecular design of polymeric materials. Marcel Dekker, New York, p 109
6. Quirk RP, Kim J-S (1995) J Phy Org Chem 8:242
7. Anderson BC, Andrews GD, Arthur P Jr, Jacobson HW, Melby LR, Playtis AJ, Sharkey WH (1984) Macromolecules 14:1599
8. Asami R, Kondo Y, Takaki M (1987) In: Hogen-Esch TE, Smid J (eds) Recent Advances in Anionic Polymers. Proceedings of the International Symposium. Elsevier, New York, p 381
9. Sogah DY, Hertler WR, Webster OW, Cohen GM (1987) Macromolecules 20:1473
10. Simms JA (1991) Rubber Chem Technol 64:139
11. Dicker IB, Cohen GM, Farnham WB, Hertler WR, Laganis ED, Sogah DY (1990) Macromolecules 23:4034
12. Brittain WJ, Dicker IB (1989) Macromolecules 22:1054
13. Reetz MT, Hutte S, Goddard R (1995) J Phys Org Chem 8:231
14. Hertler WR, Sogah DY, Webster OW, Trost BM (1984) Macromolecules 17:1415
15. Zhuang R, Muller AHE (1995) Macromolecules 28:8043
16. Hertler WR, RajanBabu TV, Overnal DW, Reddy GS, Sogah DY (1988) J Am Chem Soc 110:5841
17. Sogah DY, Webster OW (1986) Macromolecules 19:1775
18. Hirabayashi T, Itoh T (1988) Polym J 20:1041
19. Martin DT, Bywater S (1992) Makromol Chem 193:1011
20. Zou YXH, Pan R (1990) Chin Chem Lett 1:265
21. Miller J, Jenkins AD, Tsartolia E, Watson DRM, Stejskal J, Kratochvil P (1988) Polym Bull 20:247
22. Corbin DR, Sormani PME (1992) U.S. Patent 5,162,467 (to DuPont)
23. Berl V, Helmchen G, Preston S (1994) Tetrahedron Lett 35:233
24. Mei M, Muller AHE (1987) Makromol Chem Rapid Commun 8:99
25. Dixon DA, Hertler WR, Chase DB, Farnham WB, Davidson F (1988) Inorg Chem 27:4012
26. Sitz H-D, Bandermann F (1987) Group transfer of methyl methacrylate with basic catalysts In: Fontanille M, Guyot M (eds) Recent advances in mechanistic aspects of polymerization. Reidel, p 41

27. Seebach D (1988) Angew Chem Int Ed Engl 100:1624
28. Hertler WR (1994) Macromol Symp 88:55
29. Schneider LV, Dicker IB (1988) U.S. Patent 4,736,003 (to DuPont)
30. Brittain WJ (1988) J Am Chem Soc 110:7440
31. Vamvakaki M (U Crete), Patrickios CS (U Cyprus), Webster OW unpublished results. MMA, MTS, TMS benzoate (made by reaction of MTS with benzoic acid) and TBA benzoate were combined in THF (25% solution) at molar ratios of 20/1/5/0.02. The polymerization was 65% complete in 2.5 h (GPC), Mn 3300, MWD 1.07. After 24 h, Mn 4900, MWD 1.07. With no TMS benzoate the polymerization was complete in 10 min, Mn 2600, MWD 1.37
32. Uschida S (present location Tokyo Inst Tech), Webster OW unpublished results. MMA, MTS, TMS Ac, and TBA Ac were combined in THF (20% solution) at a molar ratio of 20/1/5/0.02. The polymerization was complete in 30 min (NMR). Mn 1632, MWD 1.17. Without the TMS Ac the THF refluxed and the polymerization was over in less than 1 min
33. Farnham WB, Sogah DY (1986) Polym Prepr Am Chem Soc Div Polym Chem 27:167
34. Sogah DY, Farnham WB (1985) In: Sakurai (ed) Organosilicon and biooganosilicon chemistry: stucture, bonding, reactivity and synthetic application. Ellis Horwood, Chichester, chap 20, p 219
35. Raucher S, Schindel BC (1987) Syn Comm 17:637
36. Webster OW, unpublished results
37. Quirk RP, Ren J (1992) Macromolecules 25:6612
38. Webster OW, Hertler WR, unpublished results
39. Haddleton DM, Crossman MC, Hunt KH, Topping C, Waterson C, Suddaby KG (1997) Macromolecules 30:3992
40. Jenkins AD, Tsartolla E, Walton DRM, Stejskal J, Kratochvil P (1988) Polym Bull 20:97
41. Muller MA, Stickler M (1986) Makromol Chem Rapid Commun 7:575
42. Fox TG, Schnecko HW (1962) Polymer 3:575
43. Wei Y, Wenk GE (1987) Polym Prepr Am Chem Soc Div Polym Chem 28:252
44. Gabor AH, Ober CK (1996) Chem Mater 8:2272
45. Rannard SP, Bilingham NC, Armes SP, Mykytiuk J (1993) Eur Polym J 29:407
46. Patrickios CS, Hertler WR, Abbott NL, Hatton TA (1994) Macromolecules 27:930
47. Mai PM, Muller AHE (1987) Makromol Chem Rapid Commun 8:247
48. Muller AHE (1990) Makromol Chem Makromol Shymp 32:87
49. Schubert W, Sitz H-D, Bandermann F (1989) Makromol Chem 190:2193
50. Shubert W, Bandermann F (1989) Makromol Chem 190:2721
51. Sitz HD, Speilkamp HD, Bandermann F (1988) Makromol Chem 189:429
52. Spinelli HJ (1996) Prog Org Coat 27:255
53. Szwarc M (1956) Nature 178:1168
54. Miyamoto M, Sawamoto M, Higasimura T (1984) Macromolecules 17:265, 2228
55. Faust R, Kennedy J (1986) Polym Bull 17:7
56. Adachi T, Sugimoto H, Aida T, Inoue S (1993) Macromolecules 26:1238
57. Aida T, Sugimoto H, Kuroli M, Inoue S (1995) J Phys Org Chem 8:249
58. Inoue S (2000) J Polym Sci Part A Polym Chem 38:2861
59. Yasuda H, Yamamoto H, Takemoto Y, Yamashita M, Yokoto K, Miyaki S, Nakamure A (1993) Makromol Chem Macromol Symp 67:187
60. Yasuda H (2001) J Polym Sci Part A Polym Chem 39:1955
61. Boffa LS, Novak BM (1994) Macromolecules 27:6993
62. Collins S, Ward DG, Suddaby KF (1994) Macromolecules 27:7222
63. Wang J-S, Jerome R, Teyssie P (1995) J Phys Org Chem 8:208
64. Ballard DGH, Bowles RJ, Haddleton DM, Richards SN, Sellens R, Twose DL (1992) Macromolecules 25:5907
65. Reetz MT, Knauf T, Minet U, Bingel C (1988) Angew Chem Int Ed Engl 27:1373
66. Sivaram S, Dhal PK, Kashikal SP, Khisti RS, Shinde BM, Baskaran D (1991) Polym Bull 25:77

67. Webster OW (1994) J Macromol Sci Pure Appl Chem A31:927
68. Haggard RA, Lewis SN (1984) Prog Org Coatings 12:1
69. Pietzonka T, Seebach D (1993) Angew Chem Int Ed Engl 32:716
70. Baskaran D, Muller AHE (2000) Macromol Rapid Commun 21:390
71. Konigsmann H, Jungling S, Muller HE (2000) Macromol Rapid Comm 21:758
72. Zagata AP, Hogen-Esch TE (1996) Macromolecules 29:3038
73. Baskaran D, Muller AHE (1997) Macromolecules 30:1869
74. Otsu T, Yoshida M, Tazaki T (1982) Makromol Chem Rapid Commun 3:127, 3:133
75. Bledzki A, Braun D (1981) Makromol Chem 182:1047
76. Otsu T (2000) J Polym Sci Part A Polym Chem 38:2121
77. Solomon DH, Rizzardo E, Cacioli P (1985) U.S. Patent 4,581,429
78. Rizzardo E (1987) Chem Aust 54:32
79. Georges MK, Vergin RPN, KazmaierPM, Hamer GK (1993) Macromolecules 26:2987
80. Benoit D, Chaplinski V, Braslau R, Hawker CJ (1999) J Am Chem Soc 121:3904
81. Benoit D, Grimaldi S, Robin S, Finet J-P, Tordo P, Gnanou Y (2000) J Am Chem Soc 122:5929
82. Hawker CJ, Bosman AW, Harth E (2001) Chem Rev 101:3661
83. Wang J-S, Matyjaszewski K (1995) J Am Chem Soc 117:5614
84. Kato M, Kamigaito M, Sawamoto M, Higashimura T (1995) Macromolecules 28:1721
85. Percec V, Barbiou B (1995) Macromolecules 28:7970
86. Percec V, Barbiou B, Kim H-J (1998) J Am Chem Soc 120:305
87. Matyjaszewski K, Xia J (2001) Chem Rev 101:2921
88. Kamigaito M, Ando T, Sawamoto M (2002) Chem Rev 101:3689
89. Chiefari J, Chong YKB, Ercole F, Krstina J, Jeffery J, Letp T, Mayadunne RTA, Meijs GF, Moad CL, Moad G, Rizzardo E, Thang SH (1998) Macromolecules 31:5559
90. Mayadunne RTA, Rizzardo E, Chiefari J, Chong YK, Moad G, Thang SH (1999) Macromolecules 32:6977
91. Davis TP (2002) Private communication
92. Wieland PC, Raether B, Nuyken O (2001) Macromol Rapid Commun 22:700
93. Gridnev A, Ittel SD (2001) Chem Rev 101:3611
94. Gridnev A (2000) J Polym Chem Part A Polym Chem 38:1753
95. Wayland BB, Pasmik G, Mukerjee SL, Fryd M (1994) J Am Chem Soc 116:7943

Editor: Taheo Saegusa
Received: November 2002

Adv Polym Sci (2004) 167:35–79
DOI: 10.1007/b12304

The Use of Ozone in the Synthesis of New Polymers and the Modification of Polymers

Jean Jacques Robin

Laboratoire de Chimie Macromoléculaire U.M.R. C.N.R.S.
5076 Ecole Nationale Supérieure de Chimie Montpellier, 8 Rue de l'Ecole Normale,
34296 Montpellier, France
E-mail: jrobin@cit.enscm.fr

Abstract Ozone constitutes a convenient reagent to modify natural or synthetic polymers and to functionalize them with acid or hydroxyl groups. The synthesis of telechelic oligomers or copolymers starting from ozone pre-activated polymers has been explored by numerous authors. In a first part, the mechanisms of ozone attack onto polymers of different structures such as polyolefins, polydienes and polyaromatic polymers are developed and detailed. In a second part, the synthesis of well defined copolymers starting from ozonized polymers is described. Thus, the copolymerization of different monomers starting from the thermal decomposition of peroxides and hydroperoxides resulting from the ozonization of polymers, leads to different kinds of block and graft copolymers or telechelic oligomers. In a third part, the treatment of natural polymers is described since its permits to obtain new properties. At least, the chemical modification of polymer surfaces can be achieved by treating them with ozone that confers interesting surface properties.

Keywords Ozone · Grafting · Copolymers · Telechelic oligomers · Modification

List of Abbreviations

AA	Acrylic acid
But	Butadiene
E.S.C.A.	Electronic spectroscopy for chemical analysis
F.T.I.R.	Fourier transform infra red
HEA	Hydroxy ethyl acrylate
HEMA	Hydroxy ethyl methacrylate
MADAME	Dimethyl amino ethyl methacrylate
MAGLY	Glycidyl methacrylate
MMA	Methyl methacrylate
PDMS	Polydimethyl siloxane
PE	Polyethylene
PET	Polyethylene terephthalate
PCTFE	Polychlorotrifluoroethylene
PTFE	Polytetrafluoroethylene
PHFP	Polyhexafluoropropene
POOH	Hydroperoxide
POOP	Peroxide
PP	Polypropylene
PVC	Polyvinylchloride
St	Styrene
UV	Ultra violet
VAC	Vinyl acetate
VC	Vinyl chloride

1 Introduction

The chemical modification of polymers in order to obtain particular structures or to confer specific properties to materials remains a current fundamental topic of research. So, progress and development are extensive in this area. Beside the great conventional chemical ways of synthesis leading to the most widely used polymers and to engineering polymers, during the last few years new kinds of chemical reactions based on the modification of conventional polymers have emerged. Thus, new structures, that can be used as reagents in further macromolecules and materials syntheses were obtained. This area has been widely investigated by specialists in polymer chemistry. Ozone appears as a very interesting and cheap reagent.

It is interesting to note that ozone is the subject today of many discussions in relation with environmental concerns. However, this gas remains an important chemical reagent in industrial areas like the sterilization of water or the bleaching of wood pulp [1].

The composition of this gas is very complex as a result of the coexistence of several chemical species. First, ozone is most often produced by cold plasma based on the flow of air or pure oxygen through electrodes. Oxygen is then decomposed in different species: negative ions: O^-, O_2^-, O_3^-, O_4^-, positive ions: O^+, O_2^+, O_3^+, O_4^+, neutral molecules: O, O_2, O_3 which behave as excited states [2]. The plasma of oxygen is called ozone (O_3). This gas is relatively stable up to 70 °C and can decompose in ionic species or other excited species when increasing temperature, like O_2^* being the first excited state of oxygen. Ozone is a very dipolar molecule used for its oxidative properties.

Ozone, which is commonly written as O_3 for simplification, is easily produced and does not need sophisticated apparatus that could be totally unacceptable for normal use. Therefore, this gaseous mixture composed of different entities is not stable, its destruction is relatively simple and does not require specific and expensive plants.

The first uses of ozone in chemistry were mainly restrained to analytical ones like titration of double bonds [3, 4] by adding O_3 onto unsaturations followed by titration of non-reacted ozone by iodine in solution.

Later, the use of ozone has been extensively developed in the modification of polymers or in the synthesis of new entities. We will try to summarize in this chapter the main research using ozone in the field of chemistry of polymers, in their processing, and also for their recycling at the end of their life.

2 Principles of Ozonolysis of Low Molecular Compounds

Ozone acts as a very powerful oxidation agent and is capable of cleaving double bonds in a selective and fast way. Although this reagent can react with different organic groups (amines, sulfur compounds, phosphites, multiple and hetero multiple bonds, etc.), oxidation of olefins stays the most often encountered reaction, even on an industrial scale.

Many studies relative to the obtaining of oxygen-containing products, starting from a wide variety of unsaturated compounds, in different solvents, have been mainly realized by Criegee and some other authors [5–12] and the mechanism he proposed is outlined in Scheme 1.

This mechanism proceed via a peroxidic Zwitterion what is now largely accepted by all the scientific community. Product 1, an ozone-olefin adduct is a very unstable compound giving rapidly product 3, probably through intermediate 2. The Criegee intermediate 3 can lead to different structures like:

- Ozonide 5 through reaction of Zwitterion species with aldehyde or ketone 4. This reaction can also lead to polymeric peroxide.
- Dimer 6 or polymers favored when structure 4 is a ketone; This reaction can also lead to polymeric ozonide.
- Hydroperoxide 7 through reaction of Zwitterion 3 with protic, nucleophilic molecule like ROH, H_2O, R-COO, NHR.

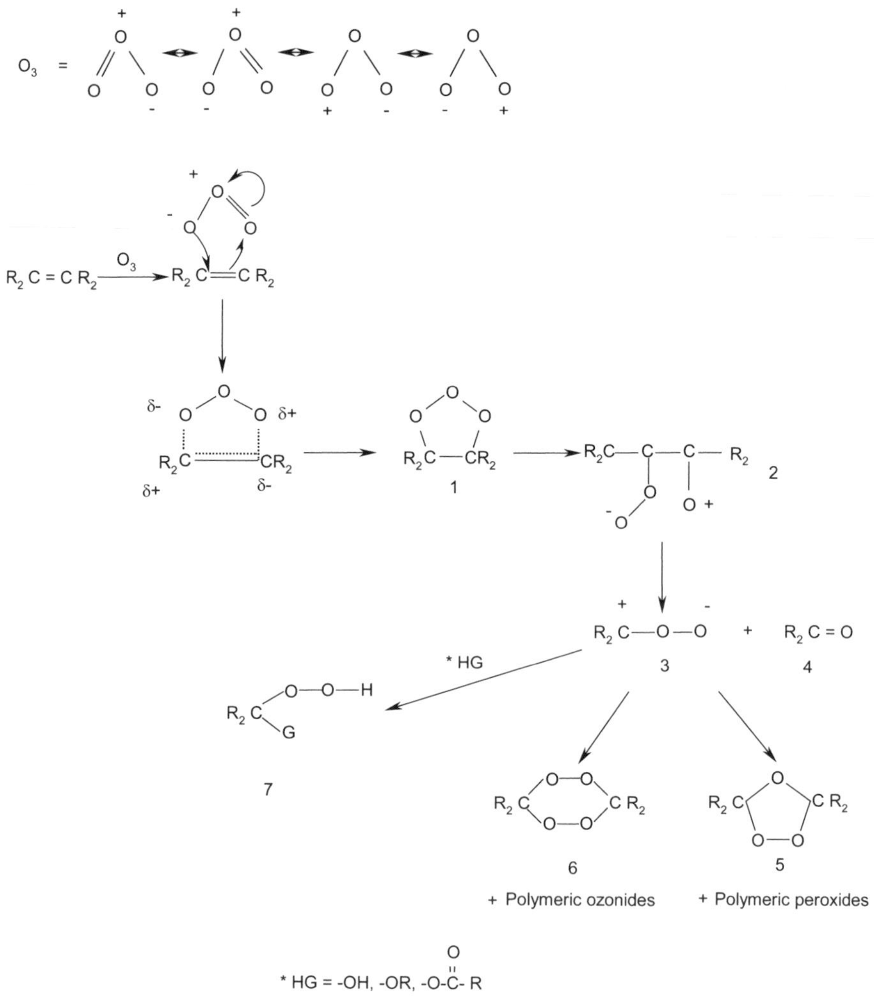

$$O_3 = \overset{+}{O}(=O)-O^- \leftrightarrow \overset{+}{O}(-O^-)=O \leftrightarrow \ldots$$

Scheme 1

* HG = -OH, -OR, -O-C(=O)-R

It must be noted here that ketone or aldehyde, diperoxides and peroxide oligomers are obtained in non participating solvents whereas alkoxy or alkyl hydroperoxides result from ozonolysis conducted in participating solvents [13].

In some cases, some abnormal products have been observed for example in the ozonolysis of allylic compounds [14] such as $CH_3CH{=}CH\text{-}CH_2OH$ where formation of HCOOH is identified.

Generally, the ozonolysis reaction shows best yields for olefins with few or small substituents. Several authors tried to define a general rule concerning the stereochemistry of this reaction without any success. The cleavage of double bonds with ozone differs greatly according to the electronic and ster-

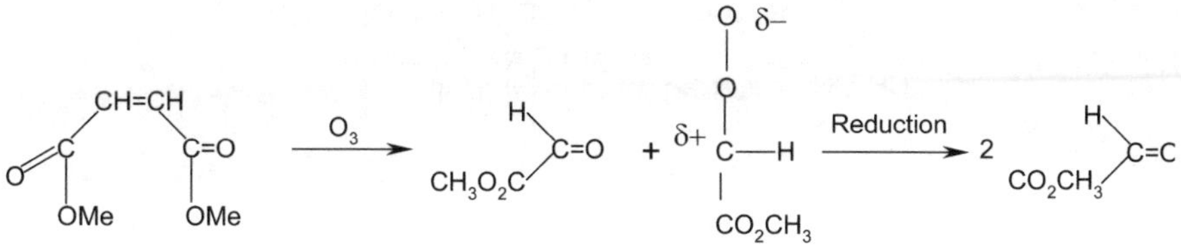
CH=CH
C=O
OMe
OMe
O₃
H
C=O
CH₃O₂C
+
δ+
O δ–
O
C—H
CO₂CH₃
Reduction
2
H
C=C
CO₂CH₃

Scheme 2

ic effect and each case has to be considered separately. Nevertheless, *cis* double bonds show higher reactivity than *trans* double bonds. These aspects of

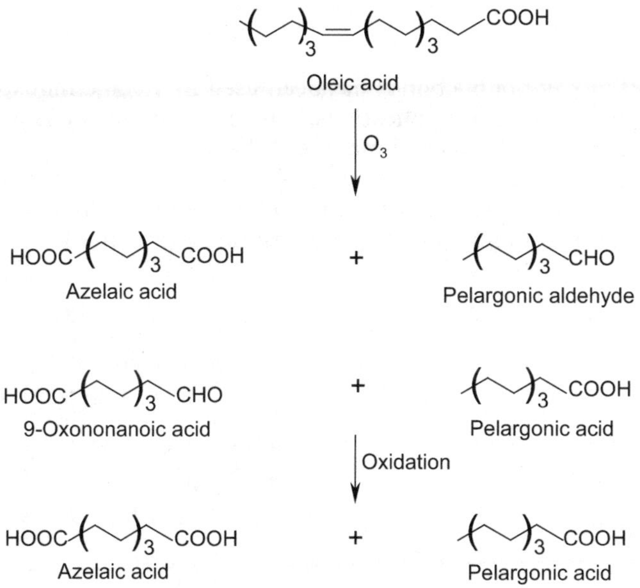
COOH
Oleic acid
O₃
HOOC COOH
Azelaic acid
+
CHO
Pelargonic aldehyde
HOOC CHO
9-Oxononanoic acid
+
COOH
Pelargonic acid
Oxidation
HOOC COOH
Azelaic acid
+
COOH
Pelargonic acid

Scheme 3

the question have been studied in details by Criegee [9–11] and other authors [15, 26] and are summed up in Bailey's review [13, 27].

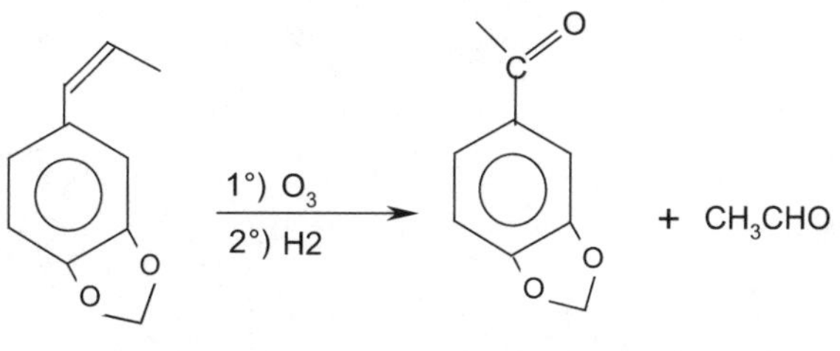
1°) O₃
2°) H2
+ CH₃CHO

Scheme 4

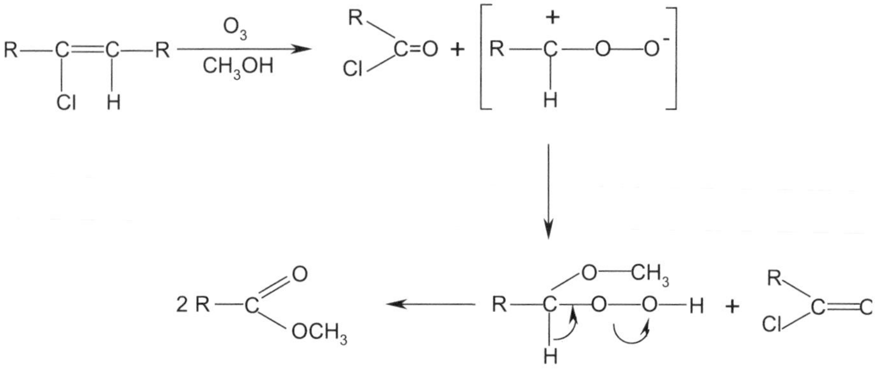

Scheme 5

The reactivity of double bonds has been used in organic industrial chemistry and pharmaceutical applications. For example, glyoxylic acid is produced by ozonolysis of dimethyl maleate (Scheme 2).

Similarly, azelaic acid (1.9-nonanedioic acid) is synthesized from oleic acid (9-octadecenoic acid) (Scheme 3) and heliatropine, which a widely used fragrance, can be produced starting from isosafrol (Scheme 4).

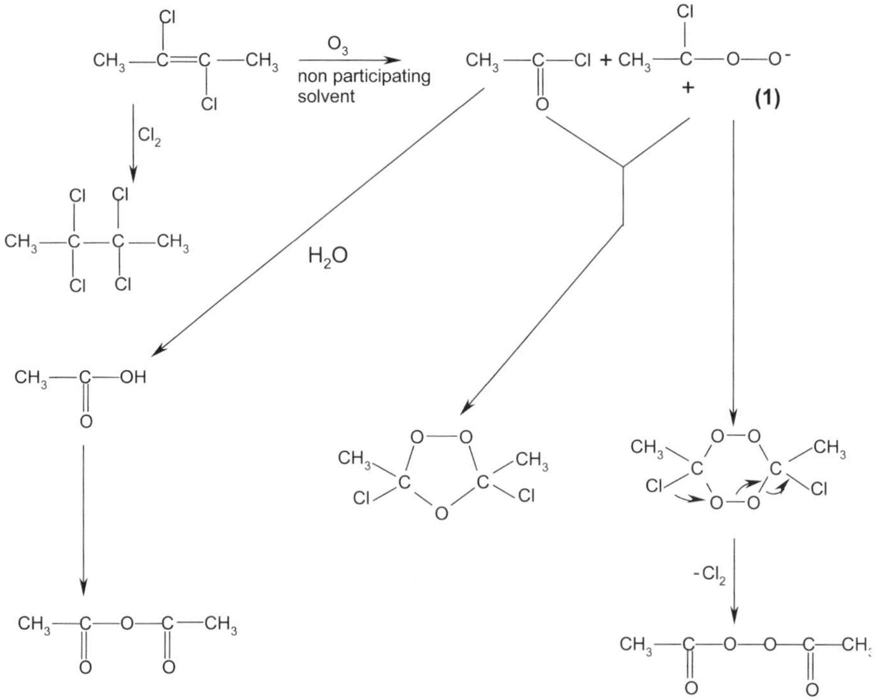

Scheme 6

Likewise, haloethylenes have been reacted with ozone and Griesbaum has shown that this type of olefinic compounds leads to esters in participating solvents [28] (Scheme 5).

1,2-Dihaloethylenes have also been studied and some surprising products were obtained, showing that in these cases, mechanisms are more complex [29, 30] (Scheme 6).

The authors identified the epoxide CH_3Cl-COC-CH_3Cl that they attributed to the addition of carbonyl oxide (1) to the olefin, which has not been clearly demonstrated [31, 32]. The epoxide formation has also been observed in the case of the ozonization of tetrachloroethylene [33–35] which yielded trichloroacetyl chloride (CCl_3-CO-Cl) and phosgene ($Cl_2C{=}O$) (Scheme 7) and in the case of a highly hindered olefin like compound (Scheme 8).

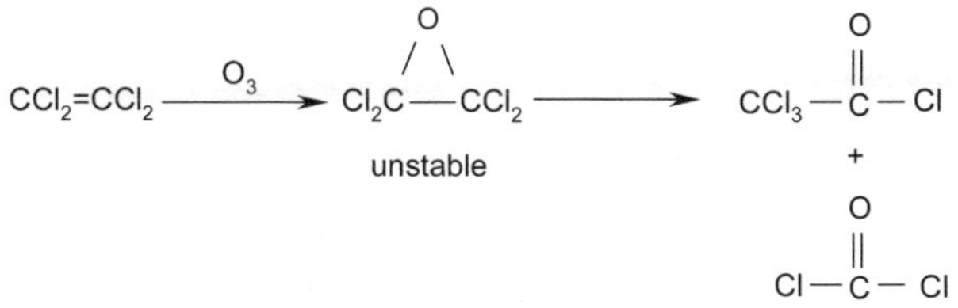

$$CCl_2{=}CCl_2 \xrightarrow{O_3} \underset{\text{unstable}}{Cl_2C\overset{O}{\frown}CCl_2} \longrightarrow CCl_3-\overset{O}{\overset{\|}{C}}-Cl \; + \; Cl-\overset{O}{\overset{\|}{C}}-Cl$$

Scheme 7

Finally, dienes such as 1,3-butadiene react easily with ozone in gas-phase reaction which gives more complex mechanisms [36, 37] than in a liquid-phase one. The proposed scheme is shown in Scheme 9.

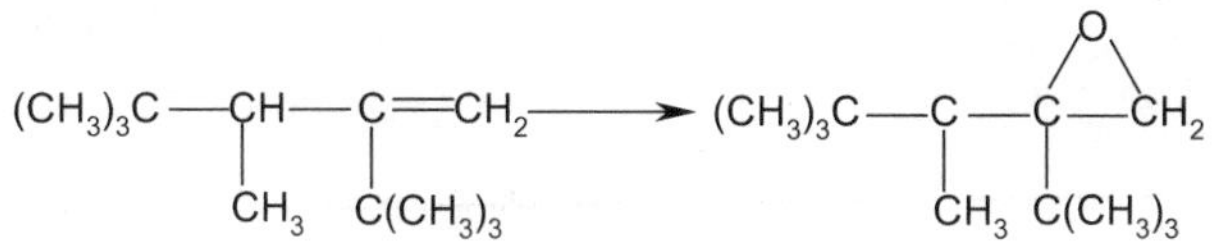

$$(CH_3)_3C-\underset{CH_3}{\underset{|}{CH}}-\underset{C(CH_3)_3}{\underset{|}{C}}{=}CH_2 \longrightarrow (CH_3)_3C-\underset{CH_3}{\underset{|}{C}}-\underset{C(CH_3)_3}{\underset{|}{C}}\overset{O}{\frown}CH_2$$

Scheme 8

Although the ozonization of double bonds has been most widely studied, other compounds can react with O_3 and a relationship taking into account rates of reaction has been established [38] (Scheme 10).

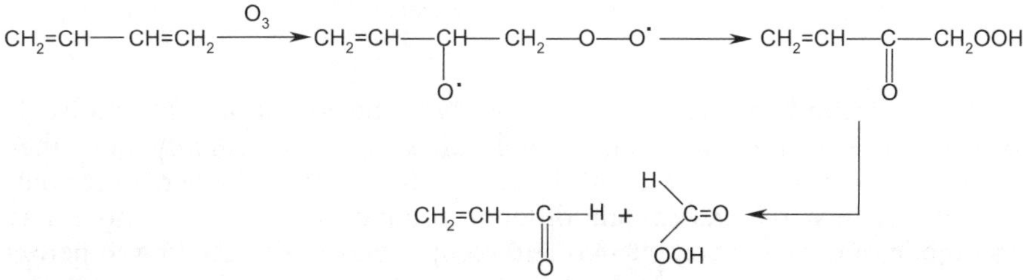

$$CH_2{=}CH-CH{=}CH_2 \xrightarrow{O_3} CH_2{=}CH-\underset{O^{\bullet}}{\underset{|}{CH}}-CH_2-O-O^{\bullet} \longrightarrow CH_2{=}CH-\underset{O}{\underset{\|}{C}}-CH_2OOH$$

$$\longrightarrow CH_2{=}CH-\underset{O}{\underset{\|}{C}}-H + \begin{matrix} H \\ \quad C{=}O \\ OOH \end{matrix}$$

Scheme 9

Criegee [39] proposed a mechanism of the ozonization of acetylenic compounds as shown in Scheme 11.

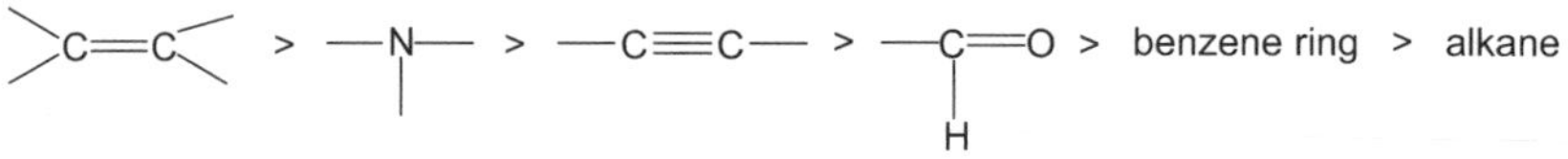

>C=C< > —N— > —C≡C— > —C(H)=O > benzene ring > alkane

Scheme 10

As it can be seen, the ozonization of alcyne gives complex mixtures of organic compounds and this type of reaction is not selective.

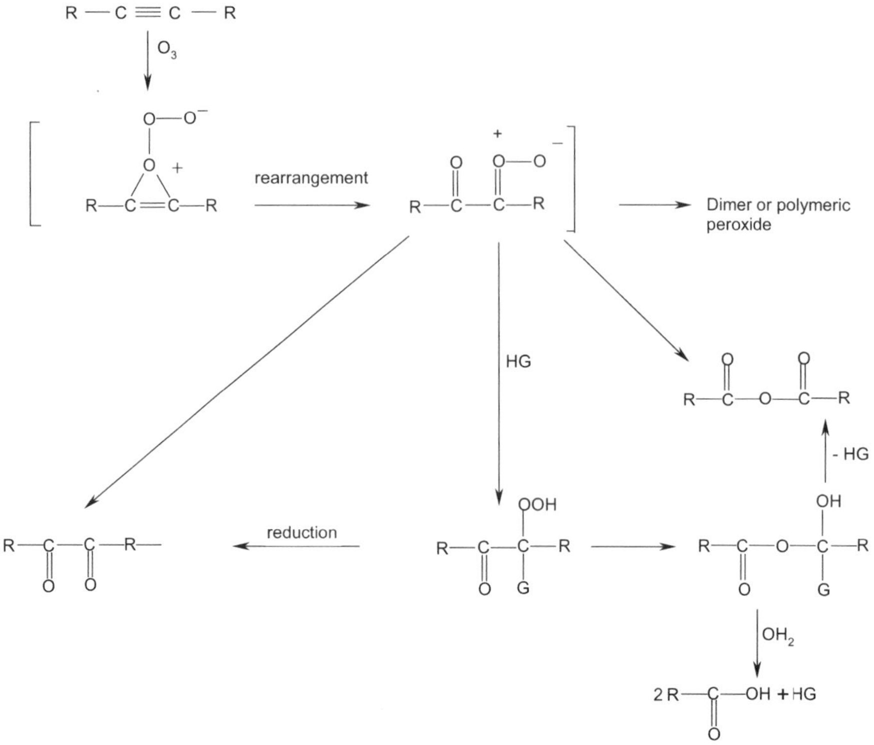

R—C≡C—R
O_3
rearrangement
Dimer or polymeric peroxide
HG
- HG
reduction
OH_2
2 R—C(=O)—OH + HG

with G = OH, R'O, R'—COO...

Scheme 11

In the case of benzene and substituted benzene, attack of a first molecule of ozone is slow but the resulting molecule reacts more rapidly since they belong to the group of olefins. At the end of the reaction, the peroxidic compounds are unstable and contain different products coming from more than one mode of ozone attack [40–43] and competing reaction leading to peroxidized and non-peroxidized structures like polymeric compounds. Harries and Wibaut [44–51] proposed the production of triozonides (Scheme 12) whereas Bailey [52] suggested the scheme shown in Scheme 13.

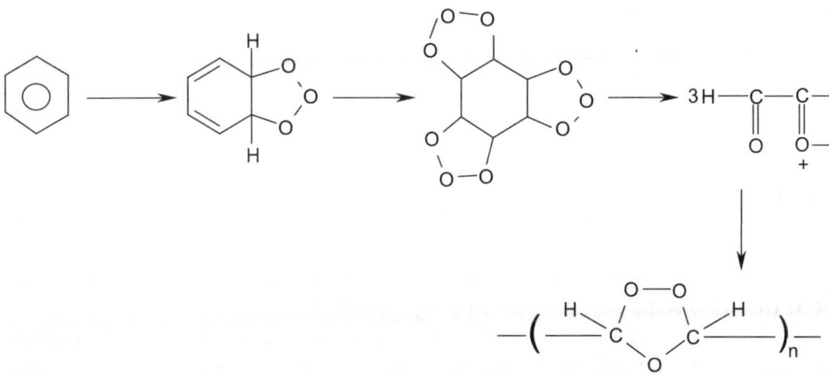

Scheme 12

Moreover, when ozonizations are performed in acids such as acetic acid or formic acid [53, 54] acyloxyalkyl hydroperoxides are formed (Scheme 14.

Other simple compounds like alcohols, aldehydes, acids, ketones have been oxidized using ozone.

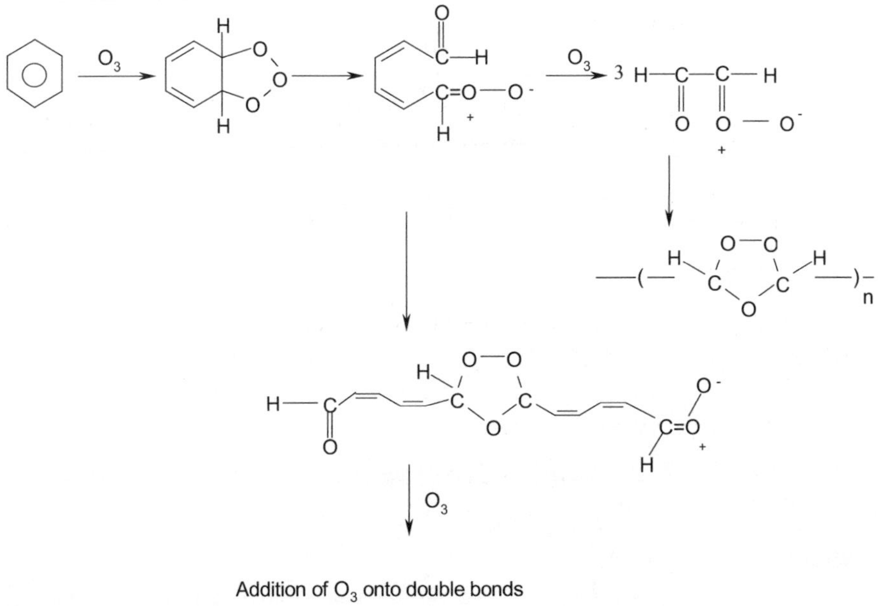

Addition of O_3 onto double bonds

Scheme 13

First, primary alcohols are transformed in their corresponding aldehydes and acids with formation of hydrogen peroxide whereas secondary alcohols give formaldehyde [55–58] (Scheme 15) where CH_3OOH will give in a further step methanol and formaldehyde.

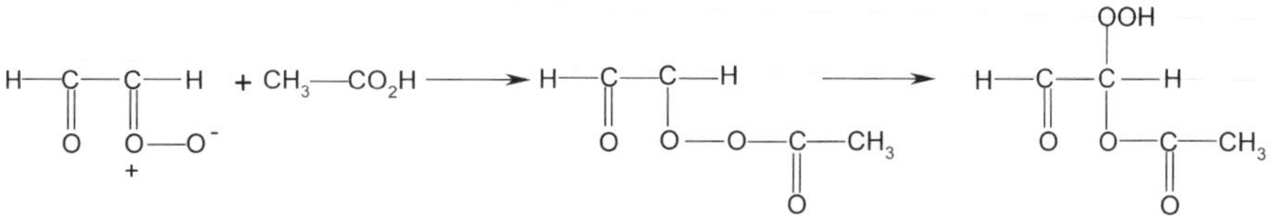

Scheme 14

Aldehydes are easily oxidized in their corresponding acid homologous [59–65] but also with formation of peracids (Scheme 16).

Paillard [66] showed in his studies that acids like acetic acid are converted in peracids and this reaction has been accelerated by the use of UV light [67].

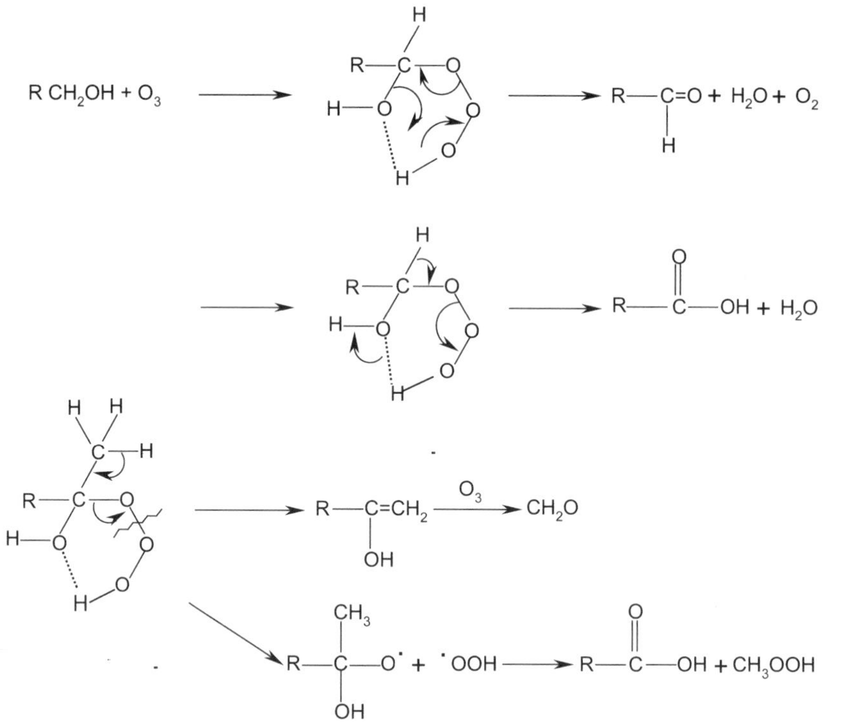

Scheme 15

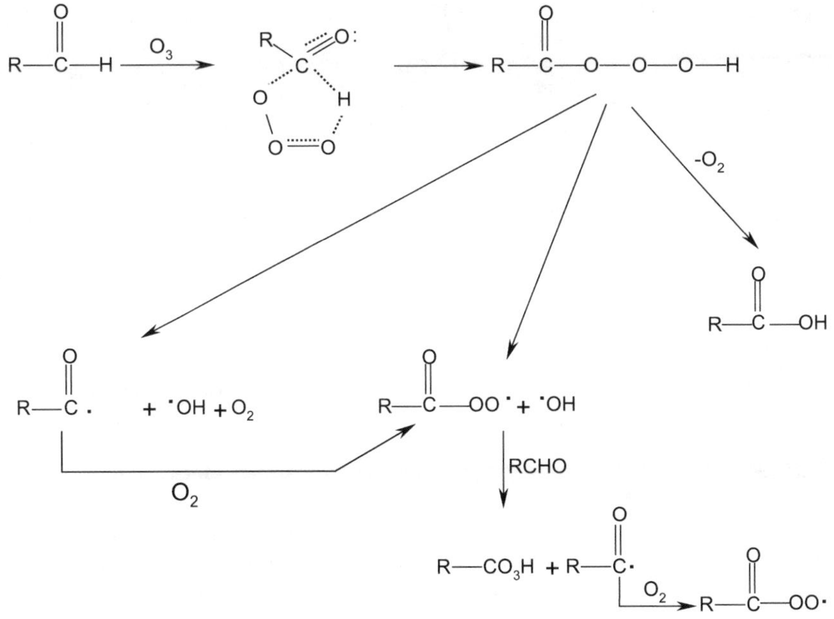

R—C(=O)—H —O_3→ [R(C=O:)(H)···O—O=O cyclic intermediate] → R—C(=O)—O—O—O—H

R—C(=O)—O—O—O—H —$-O_2$→ R—C(=O)—OH

R—C(=O)—O—O—O—H → R—Ċ(=O) + $^{\bullet}OH$ + O_2

R—C(=O)—O—O—O—H → R—C(=O)—OO$^{\bullet}$ + $^{\bullet}OH$

R—Ċ(=O) —O_2→ R—C(=O)—OO$^{\bullet}$

R—C(=O)—OO$^{\bullet}$ —RCHO→ R—CO_3H + R—Ċ(=O)

R—Ċ(=O) —O_2→ R—C(=O)—OO·

Scheme 16

In the case of ketones, the attack of O_3 is observed but is very slow and mainly occurs on the methylene group in alpha position of the carbonyl group [68–72] (Schemes 17 and 18).

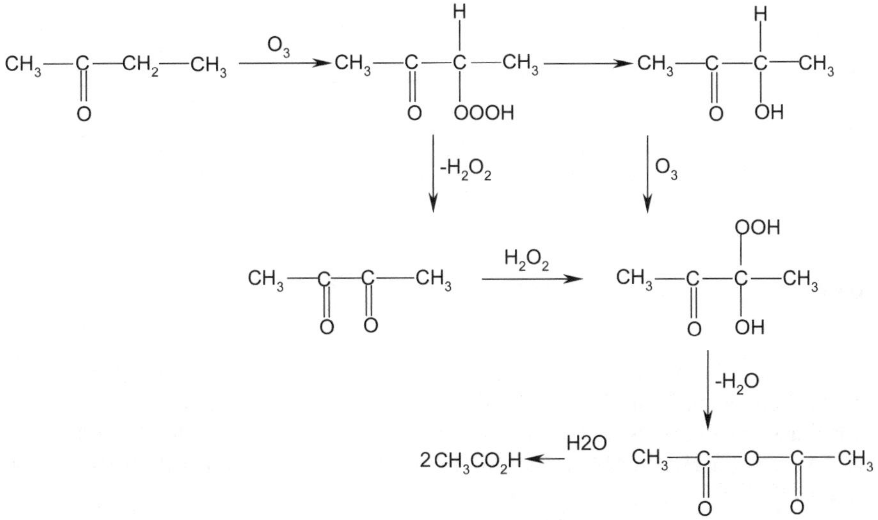

CH_3—C(=O)—CH_2—CH_3 —O_3→ CH_3—C(=O)—C(H)(OOOH)—CH_3 → CH_3—C(=O)—C(H)(OH)—CH_3

CH_3—C(=O)—C(H)(OOOH)—CH_3 —$-H_2O_2$→ CH_3—C(=O)—C(=O)—CH_3

CH_3—C(=O)—C(H)(OH)—CH_3 —O_3→ CH_3—C(=O)—C(OOH)(OH)—CH_3

CH_3—C(=O)—C(=O)—CH_3 —H_2O_2→ CH_3—C(=O)—C(OOH)(OH)—CH_3

CH_3—C(=O)—C(OOH)(OH)—CH_3 —$-H_2O$→ CH_3—C(=O)—O—C(=O)—CH_3

CH_3—C(=O)—O—C(=O)—CH_3 —H2O→ 2 CH_3CO_2H

Scheme 17

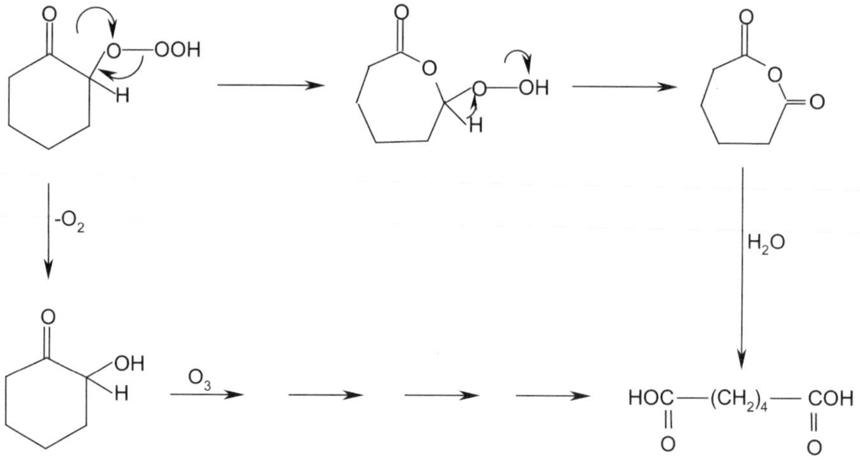

Scheme 18

At last, some workers focused their studies on organosilicon compounds [73–82]. Schemes 19 and 20 summarize the action of ozone on silanes, tetra-

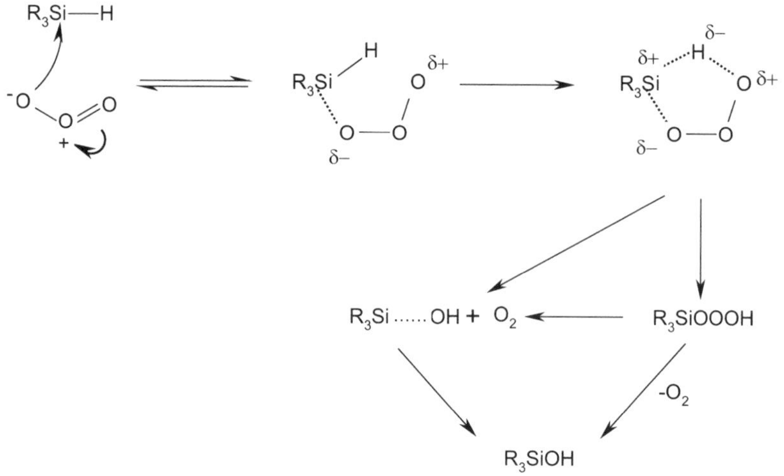

Scheme 19

substituted silanes and onto Si-Si bond [83]. It has to be noted that the ease of bond cleavage is Si-H>Si-OR>Si-OH>Si-R, and Si-Si>Si-R, and Si-Ar>Si-R and also Si-C_2H_5>Si-CH_3. So, ozonization of $(C_2H_5)_3SiCH_3$ leads exclusively to $CH_3(C_2H_5)_2Si$-O-$Si(C_2H_5)_2CH_3$ and that of $(CH_3)_3SiC_6H_5$ gives $(CH_3)_3SiOSi(CH_3)_3$.

Scheme 20

3 Study of Reaction Mechanisms of Ozone Attack Onto Polymers

3.1 Reactions with Polyolefins

The action of ozone onto polymers leads to numerous chemical modification [84] and has been studied by numerous authors, polyolefins being the most often investigated polymers. Oueslati et al. [85, 86] focused their work on polypropylene and showed that ozone leads rapidly to formation of oxygenated functions and that a chemical reaction occurred at the surface of the polymer as well as in the depth of material. So, a gradient in composition is observed. On the basis of Carlsson [87] and Razumovski's researches [88], Oueslati et al. demonstrated that the main functions appearing during treatment are unsaturated compounds, ketones, aldehydes, acids, esters, hydroxyl groups, peroxides and hydroperoxides, and finally, γ-lactones. The main mechanism is illustrated in Scheme 21.

This complicated scheme shows that the species $R^{\cdot}$, $HO^{\cdot}$, and $HOO^{\cdot}$ are responsible for the further abstraction of hydrogen from polypropylene leading to alcohol functions by abstraction of hydrogen from a neighboring chain of polymer, but also and mainly, to the β-scissions. Moreover, the radicals obtained can react with oxygen giving hydroperoxides. On the other

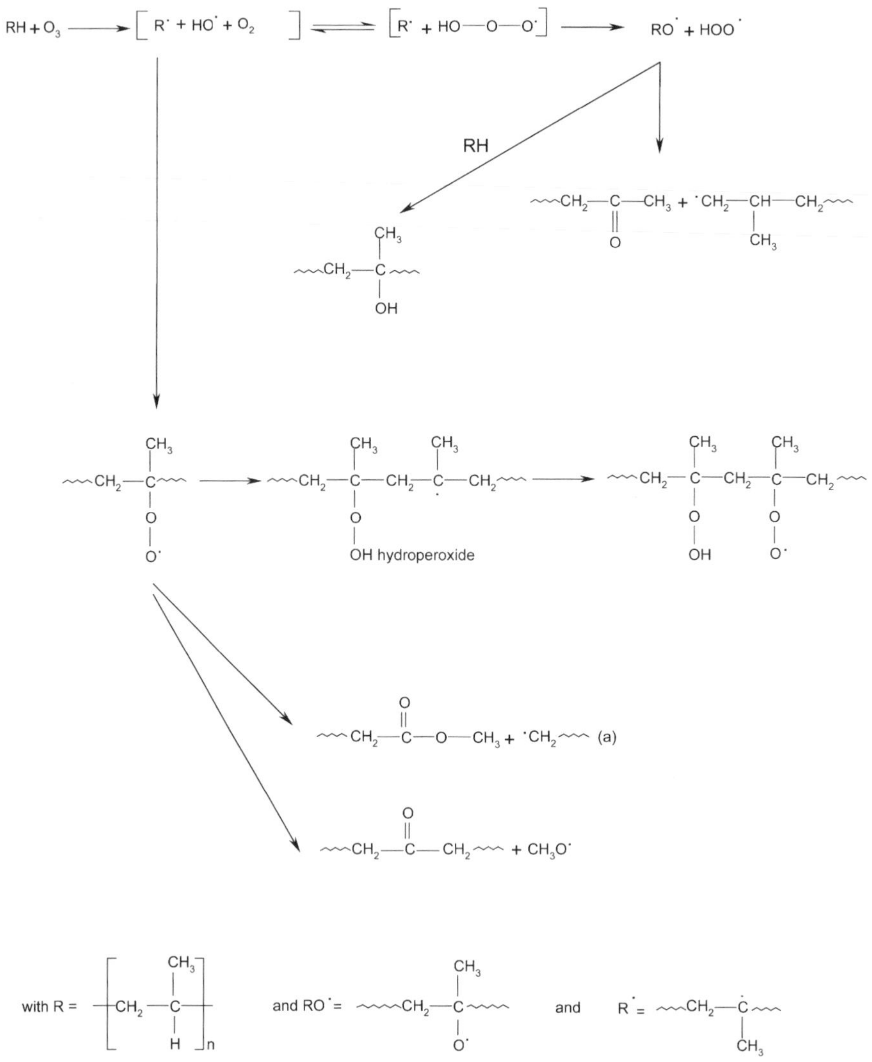

$RH + O_3 \longrightarrow [R^{\cdot} + HO^{\cdot} + O_2] \rightleftharpoons [R^{\cdot} + HO{-}O{-}O^{\cdot}] \longrightarrow RO^{\cdot} + HOO^{\cdot}$

Scheme 21

hand, intramolecular rearrangement produces different species leading to esters and ketones.

We have to mention that, in a minor proportion, the ozone attack can occur on a secondary carbon (Scheme 22).

Compounds (a) (Scheme 21) and (b) (Scheme 22) can generate alcohol functions by usual combination of hydroxyl-radical and carbon-radical (Scheme 23).

The formation of γ-lactones can be explained by intramolecular rearrangements as shown in Scheme 24.

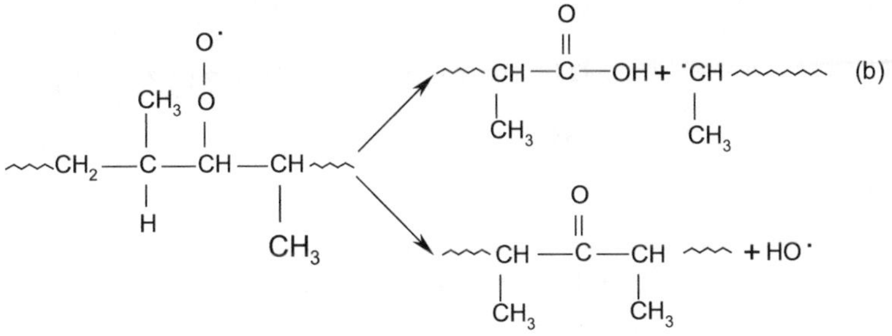

Scheme 22

At last, unsaturated compounds come from either intramolecular water elimination (Scheme 25) or normal rearrangements (Scheme 26).

$$R^{\cdot} + HO^{\cdot} \longrightarrow ROH$$

Scheme 23

In this case, the last unsaturated compound (c) is sensitive to different radical attacks coming from all radicals existing in the medium (Scheme 27).

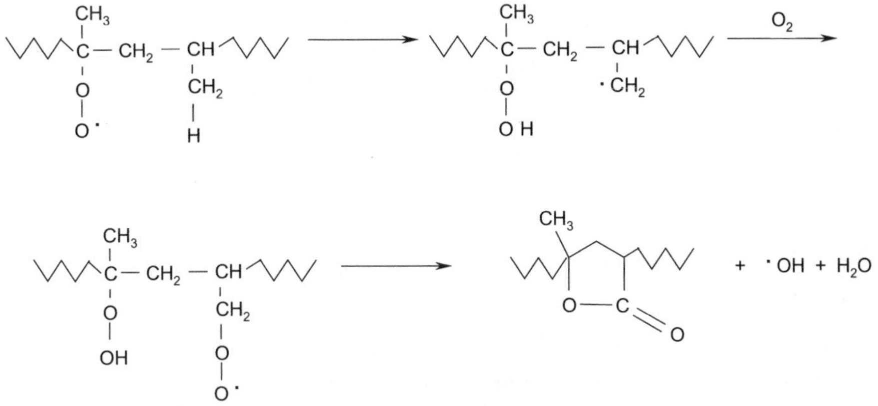

Scheme 24

The whole reaction scheme leads to numerous chain scissions, the major mechanism being the β-scission as has been demonstrated by A. Michel et al. [89]. These authors showed that the formation of carbonyl functions is directly related to chain scissions and that peroxidic species are mainly hydroperoxides rather than peroxides inserted in the macromolecular chain.

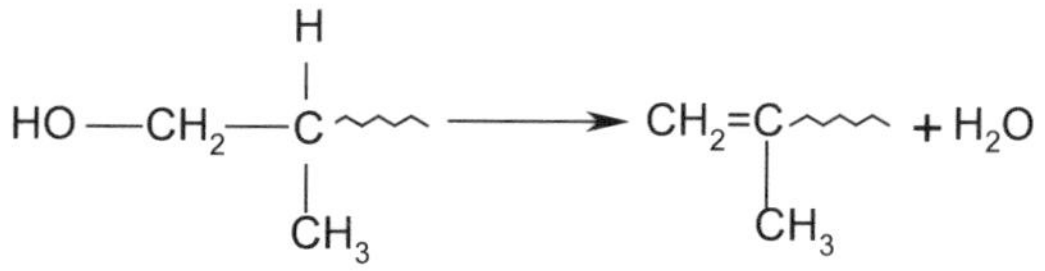

$$HO-CH_2-C(H)(CH_3)\sim\sim \longrightarrow CH_2=C(CH_3)\sim\sim + H_2O$$

Scheme 25

The ozonization of polyethylene constitutes a case derived from ozonization of polypropylene since this polymer presents branchings involving tertiary carbons and also double bonds coming from the type of polymer synthesis. Concerning the mechanisms occurring during the different steps of ozonization, the schemes described for PP are valid for PE. Razumovski et al. [88] must be mentioned here, as they were pioneers in the elucidation of the peroxidation process of polymers and proposed the equation in the case of PE shown in Scheme 28.

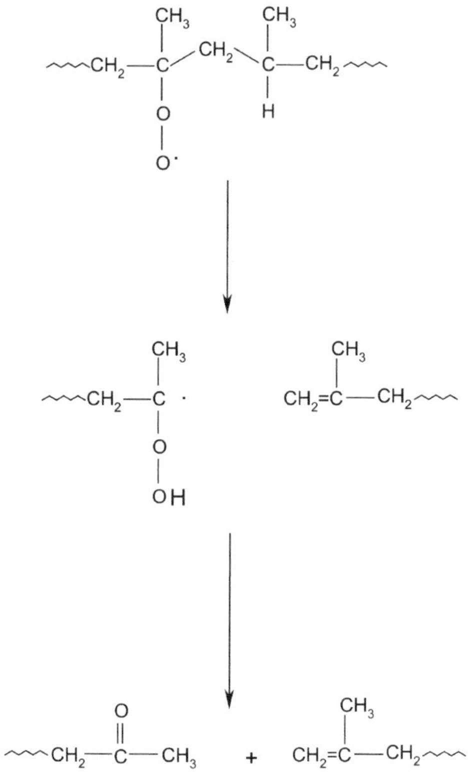

$$\sim\sim CH_2-C(CH_3)(O-O^\cdot)-CH_2-C(CH_3)(H)-CH_2\sim\sim$$

$$\downarrow$$

$$\sim\sim CH_2-C^\cdot(CH_3)(O-OH) \qquad CH_2=C(CH_3)-CH_2\sim\sim$$

$$\downarrow$$

$$\sim\sim CH_2-C(=O)-CH_3 \quad + \quad CH_2=C(CH_3)-CH_2\sim\sim$$

Scheme 26

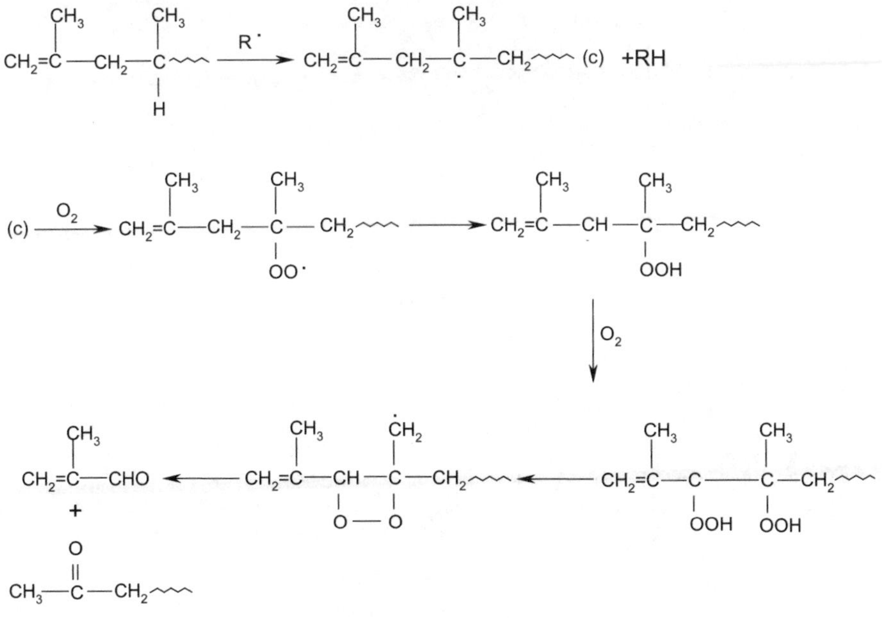

$$CH_2{=}C(CH_3){-}CH_2{-}CH(CH_3)\sim\sim \xrightarrow{R^{\cdot}} CH_2{=}C(CH_3){-}CH_2{-}\dot{C}(CH_3){-}CH_2\sim\sim \;(c)\; + RH$$

$$(c) \xrightarrow{O_2} CH_2{=}C(CH_3){-}CH_2{-}C(CH_3)(OO^{\cdot}){-}CH_2\sim\sim \longrightarrow CH_2{=}C(CH_3){-}CH{-}C(CH_3)(OOH){-}CH_2\sim\sim$$

$$\xrightarrow{O_2} CH_2{=}C(CH_3){-}C(OOH){-}C(CH_3)(OOH){-}CH_2\sim\sim \longrightarrow CH_2{=}C(CH_3){-}CH{-}C(\dot{C}H_2)(O{-}O){-}CH_2\sim\sim \longrightarrow CH_2{=}C(CH_3){-}CHO + CH_3{-}C(=O){-}CH_2\sim\sim$$

Scheme 27

PE/PP blends have also been studied in terms of stability to oxidation. Livanova shown that oxidative degradation reaches a maximum when PP is in an isotropic structure and minimum for an oriented one [90].

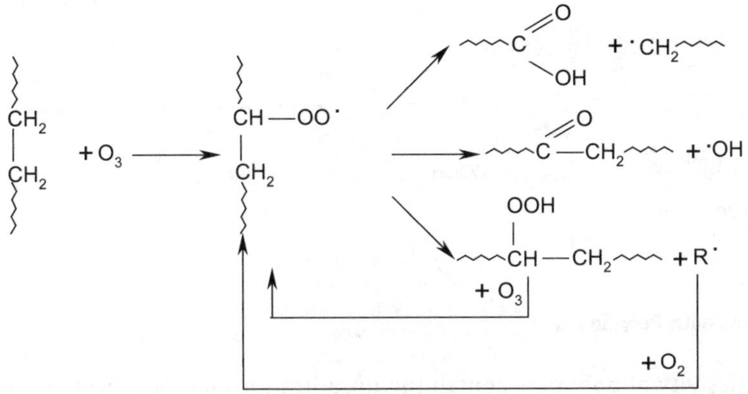

$$\sim CH_2{-}CH_2\sim + O_3 \longrightarrow \sim CH(OO^{\cdot}){-}CH_2\sim \longrightarrow \begin{cases} \sim C(=O)OH + {}^{\cdot}CH_2\sim \\ \sim C(=O){-}CH_2\sim + {}^{\cdot}OH \\ \sim CH(OOH){-}CH_2\sim + R^{\cdot} \end{cases}$$

($+ O_3$ and $+ O_2$ return arrows back to the chain)

Scheme 28

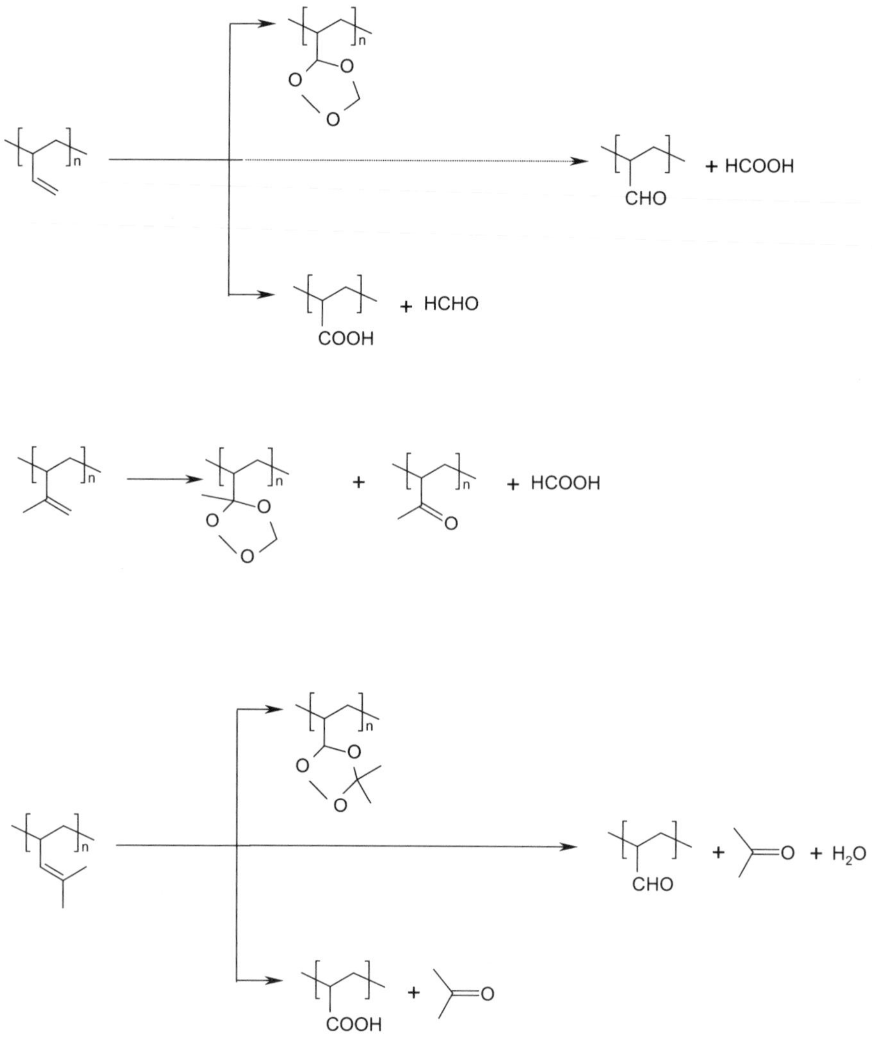

Scheme 29

3.2 Reactions with Polydienes

The reactivity of polymers containing unsaturations has also been extensively studied in terms of reactivity and structures [91, 92].

So Cataldo et al. [101] have focused their studies on polymers like 1-2 polybutadiene, 3-4 polyisoprene and poly(4-methyl-1,3 pentadiene). These polymers are very sensitive to ozone, and their own tacticity seems to have a very low influence on their reactivity. In a usual way, substituents make them more sensitive to ozone attack. For polymers containing unsaturations in their backbone, we observe a very rapid decrease in molecular weights, that is not the case for polymers presenting lateral unsaturations as pure 1-2-polybutadiene [102–104]. We observe at a very rapid decrease in molecular weights that can be correlated to the number of unsaturations in the polymer backbone. According to the reaction schemes described above, the products obtained in the experimental conditions applied by authors are mainly carbonyl groups, particularly aldehydes, acids, but also ozonides (Scheme 29).

The reactivity of unsaturations with ozone has been applied to produce structures which allow subsequent degradation of materials by ozonolysis. In this way, Peters et al. [105] prepared polyurethanes using novel unsaturated diisocyanates which can be degraded by oxidative cleavage of the double bonds.

3.3
Reactions with Aromatic Polymers

The ozonization of polymers presenting aromatic structures has been subjected to other studies, but in a less detailed way than polymers described in the last section. We have to mention here the most famous results of S.D. Razumovski and G.E. Zaikov [88] who proposed the reactions shown in Schemes 17 and 18. Their work demonstrates that the second mechanism remains the major and that, in the most often encountered cases, cycles are attacked in a second step after ring substituents. Further, experimental conditions (ozonization in solid phase or in dilute solution) leads to different products and the appearance of cross-linking is specific to solid phase ozonization.

3.3.1
Ozonide Formation

This is illustrated in Scheme 30.

3.3.2
Radical Formation

The subject of degradation of aromatic polymers, especially polystyrene remains controversial. Zaiko et al. [106] published a review focused on this theme (Scheme 31).

Radicals supported by macromolecular chains induce numerous cross-linking, which has been proved by Saito [107] and Razumovskii [108], but also favor the attack of aromatic rings as demonstrated by Kefely [109] and Peeling [110] (Scheme 32).

Saito et al. [107] investigated ozonolysis of polystyrene in carbon tetrachloride. Their results indicated the formation of carboxylic acids and,

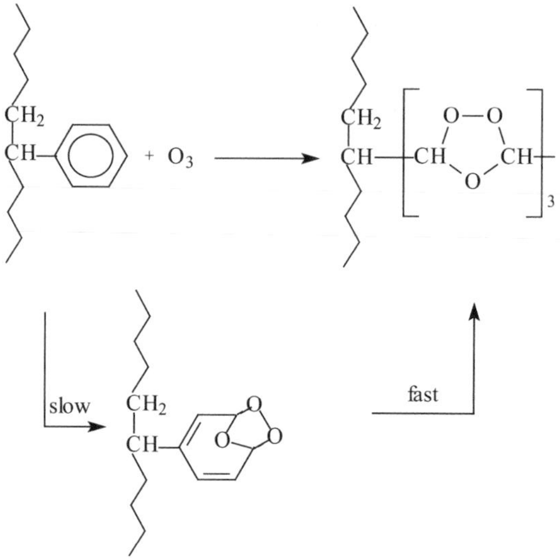

Scheme 30

above all, cross-linking. Although the tertiary carbon of PS is less sensitive to O_3 attack than that of polypropylene, it seems to be the favorite site constituting the preliminary step of the overall reaction. In the next steps, the cleavage of aromatic rings has been postulated by several authors who mentioned in addition to that, the formation of formic acid in products generated by the successive degradative reactions. Despite these different works, the real mechanisms have not been yet clearly elucidated and the ozonization of polystyrene seems to be difficult to overcome in terms of control of the obtained structures. This subject is still widely studied in order to clarify the mechanisms for which all authors do not agree [112].

3.4 Reactions with Halogenated Polymers

Among the well known polymers, the ozonolysis of PVC was subject to a great number of studies, especially by A. Michel et al. [113–115]. These authors showed that the sensitivity of this polymer depends closely at first on the temperature, and also on the structure of the starting polymer, and thus, on its way of synthesis. So, the presence of double bonds on the polymer backbone constitutes a weak point which will constitute the starting step of the degradation leading to a very quick decrease in molecular weights.

Hydrogen atoms in allylic position are favorite sites for hydroperoxidation of chains. So, this mechanism proceeds in the formation of lateral hydroperoxides, and not like for other polymers, in intramolecular peroxides. Rearrangement of chemical structures coming from ozonides are rapidly observed (Scheme 33).

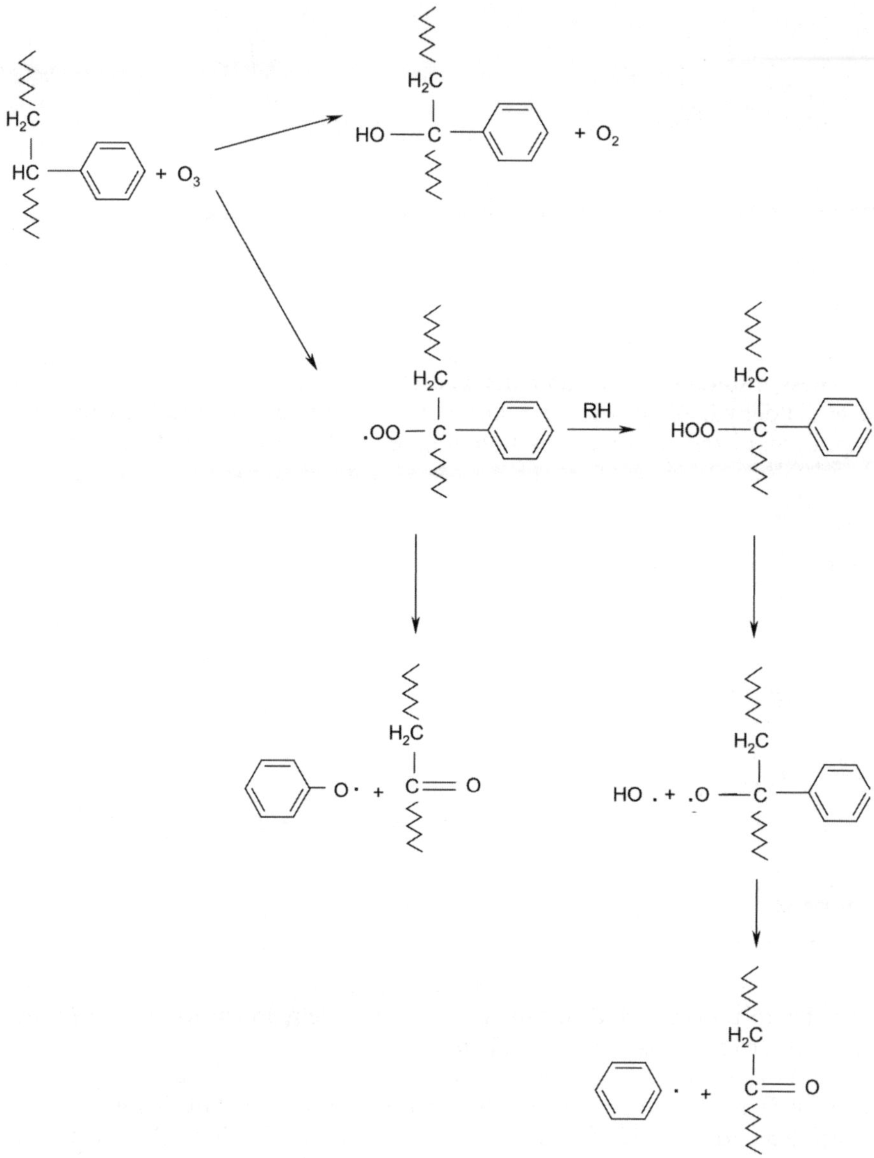

Scheme 31

In a second step, hydroperoxides lead to intramolecular rearrangements (Scheme 34).

Low molecular weight radicals lead to different oxidative products that have not been clearly characterized by authors. One hypothesis could be that reaction (b) occurs in a last step.

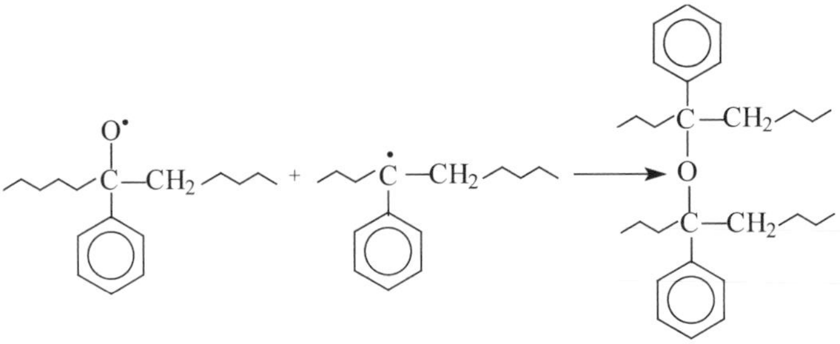

Scheme 32

These different rearrangements lead to chain scissions with formation of α-ω dialdehyde oligomers presenting very low molecular weights which are rapidly oxidized to their acid homologues and to by-products and hydro-

Scheme 33

chloric acid. Structures like those shown in Scheme 35 will lead to hydrogen transfer to polymer chains, thus giving a new dehydrochlorination producing further reactions as described above.

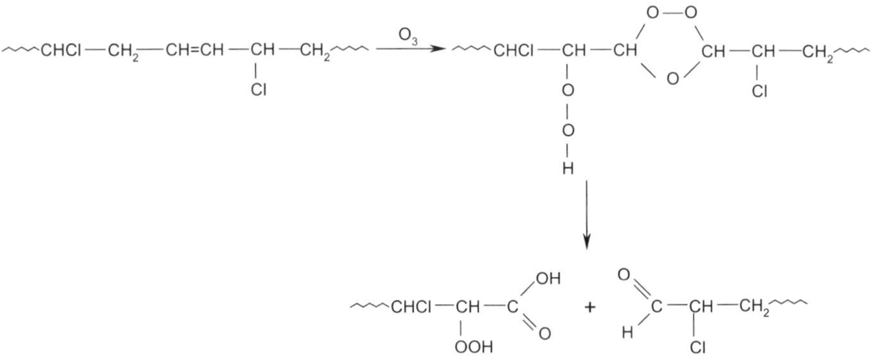

Studies performed by Michel et al. [113–115] showed that the peroxide content is in the range 0.5–40×10^{-5} mol g^{-1} and the acidity content is between 0.5 and 60 mole of acid functions per initial PVC chain. This last value is consistent with the observed decreases in molecular weights and evidenced by Landler and Lebel [116].

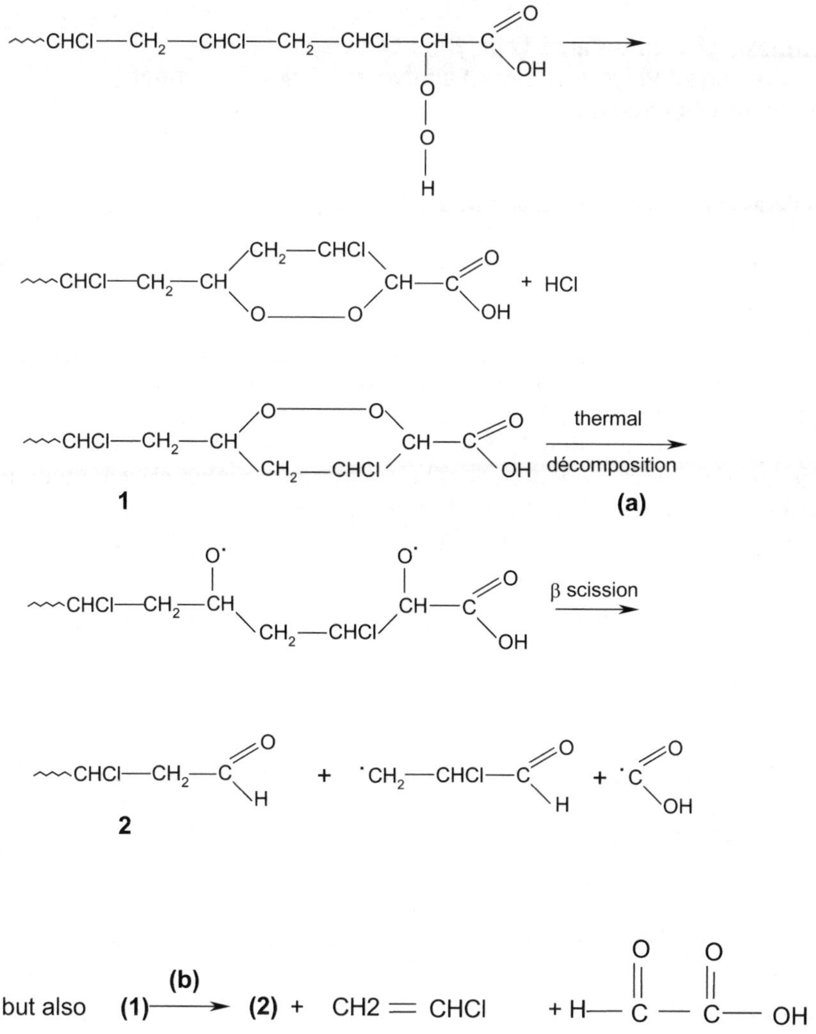

$$\sim\text{CHCl—CH}_2\text{—CHCl—CH}_2\text{—CHCl—CH(OOH)—C(=O)OH} \longrightarrow$$

$$\sim\text{CHCl—CH}_2\text{—CH}\langle\text{CH}_2\text{—CHCl}, \text{O—O}\rangle\text{CH—C(=O)OH} + \text{HCl}$$

$$\sim\text{CHCl—CH}_2\text{—CH}\langle\text{O—O}, \text{CH}_2\text{—CHCl}\rangle\text{CH—C(=O)OH} \xrightarrow[\text{décomposition}]{\text{thermal}}$$

1 **(a)**

$$\sim\text{CHCl—CH}_2\text{—CH(O}^\bullet\text{)—CH}_2\text{—CHCl—CH(O}^\bullet\text{)—C(=O)OH} \xrightarrow{\beta\ \text{scission}}$$

$$\sim\text{CHCl—CH}_2\text{—C(=O)H} + {}^\bullet\text{CH}_2\text{—CHCl—C(=O)H} + {}^\bullet\text{C(=O)OH}$$

2

but also **(1)** $\xrightarrow{\textbf{(b)}}$ **(2)** + CH2 = CHCl + H— C(=O) — C(=O) — OH

Scheme 34

$${}^\bullet\text{CH}_2\text{—CHCl—C(=O)H}$$

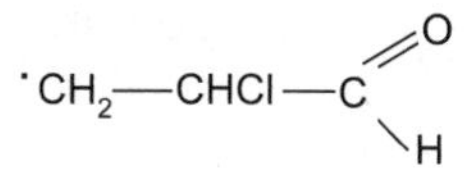

Scheme 35

4
Synthesis of Well Defined Copolymer Structures Starting from Ozonized Polymers: Block Copolymers, Graft Copolymers, Telechelic Oligomers

4.1
Synthesis of α,ω-Functional Oligomers and Block Copolymers

The synthesis of telechelic oligomers by oxidative cleavage has been extensively studied by numerous authors and Cheradame [117] reviewed the main reactions leading to telechelic polymers starting from high molecular polymers. As he showed, ozonolysis remains one of the preferred method in addition to Ruthenium tetroxide oxidation to obtain α-ω functional oligomers.

Rimmer [118] and Caffetera [119] focused their studies on the reactivity of polymers bearing double bonds in order to synthesize macroinitiators able to give block copolymers. These authors copolymerized monomers such as methyl methacrylate (MMA) or styrene (St) with 2-3 dimethyl butadiene giving copolymers presenting the structure shown in Scheme 36 with

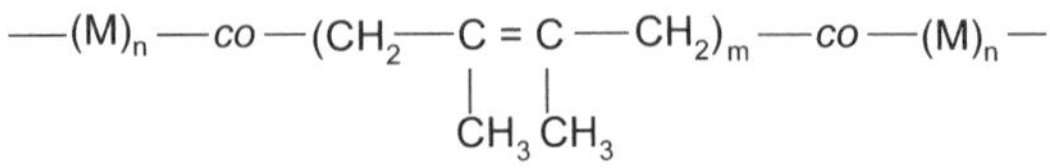

$$-(M)_n-co-(CH_2-\underset{CH_3}{\underset{|}{C}}=\underset{CH_3}{\underset{|}{C}}-CH_2)_m-co-(M)_n-$$

Scheme 36

a molar ratio of 1.6–1.8% of unsaturated monomer. In a second step, the ozonization at low temperature (–60 °C) of these copolymers lead to particular diperoxides (Scheme 37).

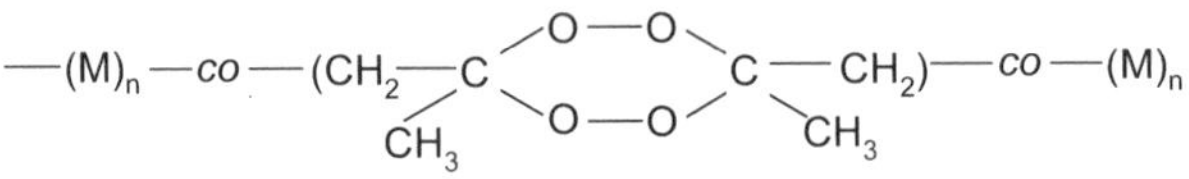

$$-(M)_n-co-(CH_2-C(CH_3)\langle{}^{O-O}_{O-O}\rangle C(CH_3)-CH_2)-co-(M)_n$$

Scheme 37

Above 60 °C this polymer decomposes, generating free radicals able to initiate copolymerization with an second monomer. In this way, the authors obtained copolymers with molecular weight of 5000 g mol^{-1}.

Model studies applied to simple molecules permitted one to propose the mechanism [119] shown in Scheme 38.

This scheme explaining the copolymerization observed has not yet been explained in kinetic aspects. Indeed, if decomposition rate of diacetyl peroxide equals 10^{-5} s^{-1} at 60 °C [120], this value does not follow the classic evolution of rate vs temperature (according to Arrhenius law). The authors suggest induced decompositions of these peroxides by $CH_3^{\cdot}$ radicals existing in the medium, and also by macroradicals coming from growing chains during polymerization of the monomer. So a great polydispersity can be observed for the obtained products.

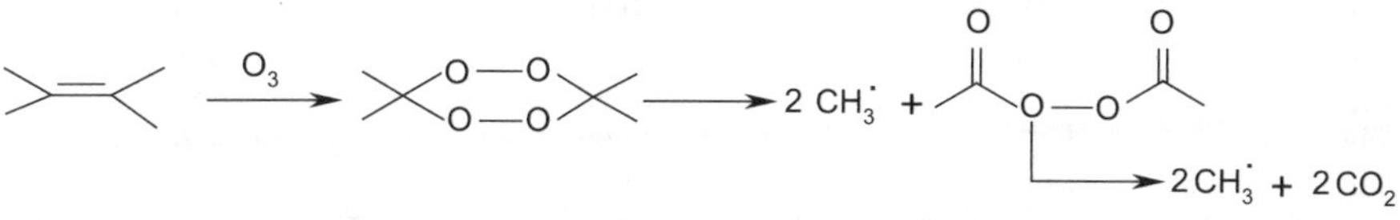

$$\mathrm{(CH_3)_2C{=}C(CH_3)_2} \xrightarrow{O_3} \text{ozonide} \longrightarrow 2\,CH_3^{\cdot} + CH_3C(O)\text{–O–O–}C(O)CH_3 \longrightarrow 2\,CH_3^{\cdot} + 2\,CO_2$$

Scheme 38

Other works have been developed by Rimmer and Ebdon [121, 122] on the subject of ozonolysis of polymers, and mainly, on unsaturated polymers. These authors activated copolymers of butadiene and methyl methacrylate P (MMA co But) by ozone, but also copolymers of butadiene and styrene P(St-But). Concerning this last copolymer, polybutadiene segments are first attacked by ozone, but the authors proved that aromatic rings are rapidly degraded, which should be avoided. They discovered that when ozonolysis is conducted in solution in *N,N′*-dialkyl amides, like dimethyl acetamide, this drawback can be avoided. Indeed, ozone reacts with substituted amides with a rate which is intermediary between that of ozone onto double bond and that of ozone onto aromatic rings. In fact, the excess of ozone in solution which did not react with double bonds of polybutadiene segments, is destroyed by reacting with *N,N′*-dimethyl acetamide. According to this method, these authors obtained, α,ω-dihydroxyl oligopolystyrene and α,ω-dialdehyde oligopolymethyl methacrylate [123] the end groups of which will later be converted by chemical modification in acid groups or sulfonate groups (sulfonation). Furthermore, polymer-supported oxidizing or reducing agents were successfully used for the chemical modification of chain ends of the ozonized polymers, mainly with the aim of improving yields and well controlling the nature of end groups.

Other authors took advantage of the use of ozone in the object of block copolymerization. Smets et al. ozonized polypropylene and studied the obtained hydroperoxides/peroxides ratios [124, 125] vs ozonization time. They show that hydroperoxide content is favored by long ozonization time (Table 1).

After a treatment at 100 °C in oxygen atmosphere, intra- and inter-molecular peroxides are formed. This phenomenon occurs to the detriment of hydroperoxides which content decreases to the benefit of peroxide content. So it becomes conceivable to adjust the reactivity of polymer towards monomer, knowing that peroxides and hydroperoxides do not present the same decomposition rate. In this way, Smets et al. [125] experimented with the copolymerization in emulsion of these activated polymers with (meth)acrylic or styrenic monomers at low temperature (35 °C) using reducing agents such as amines. During this stage, hydroperoxide functions behave as favorite sites for the initiation of monomer M. In a second stage, increasing

Table 1 Peroxide and hydroperoxide contents for ozonized PP at 22 °C [125]

Ozonization time	[POOP] mol/g of PP×10^5	[POOH] mol/g of P×10^5	% POOH
1 h	0.30	0.25	45
2 h	0.75	0.90	55
3 h	0.85	1.35	61
4 h	0.85	1.85	69
5 h	1.15	5.05	76
6 h	1.6	5.95	79

the temperature to 50 °C permits the decomposition of intra- or inter-molecular peroxide functions, thus producing copolymerization of a second monomer such as vinyl chloride (VC). We have to note here that the decomposition of hydroperoxides in RO$^·$ and HO$^·$ gives homopolymers HO-$(M)_n^·$ which can react in the second step with VC to give diblock segments and finally transfer to polymer (entity 1 as described in the scheme below). On the contrary, peroxide decomposition does not produce homopolymerization of monomers as above [125].

Based on this principle, different block and graft copolymers presenting molecular weights between 400,000 and 1,000,000 g mol^{-1} have been prepared, the proportion of the different blocks being adjusted by the initial stoichiometry of monomers.

The authors demonstrated that polyvinyl chloride segments of these graft and block copolymers are more stable than PVC homopolymers prepared according to traditional ways. These last ways lead to disproportionation reactions giving unsaturations at the ends of polymers and these ends are responsible for the poor thermal stability of the polymers. In the present method, termination occurring essentially by transfer reaction, no unsaturations are observed and the authors showed an improvement in stability of the PVC segments of about 20 °C (Scheme 39).

Other authors have been interested in the synthesis of multiblock copolymers starting from ozonization of PVC which essentially leads to α-ω acid oligomers. Michel et al. [113, 115, 126] synthesized the compounds shown in Scheme 40 using ozonization of PVC in solvents.

In all cases, authors showed that such copolymers are stable only if intramolecular peroxides resulting from ozonization have been preliminary thermally decomposed before the above reactions. Their presence in the polymer backbones leads, in a very short time, to dehydrochlorination reactions and degradation of polymers. These works have been extended to other polymers, such as chlorinated polyethylene, vinyl chloride-vinyl acetate copolymers [127].

Numerous authors took advantage of the reactivity of double bonds towards ozone to prepare α-ω functional oligomers usable in the synthesis of multiblock copolymers by copolycondensation, or in the synthesis of precursors of surfactants or ionomer resins. Results in this field of investigation are numerous, mainly in terms of industrial applications.

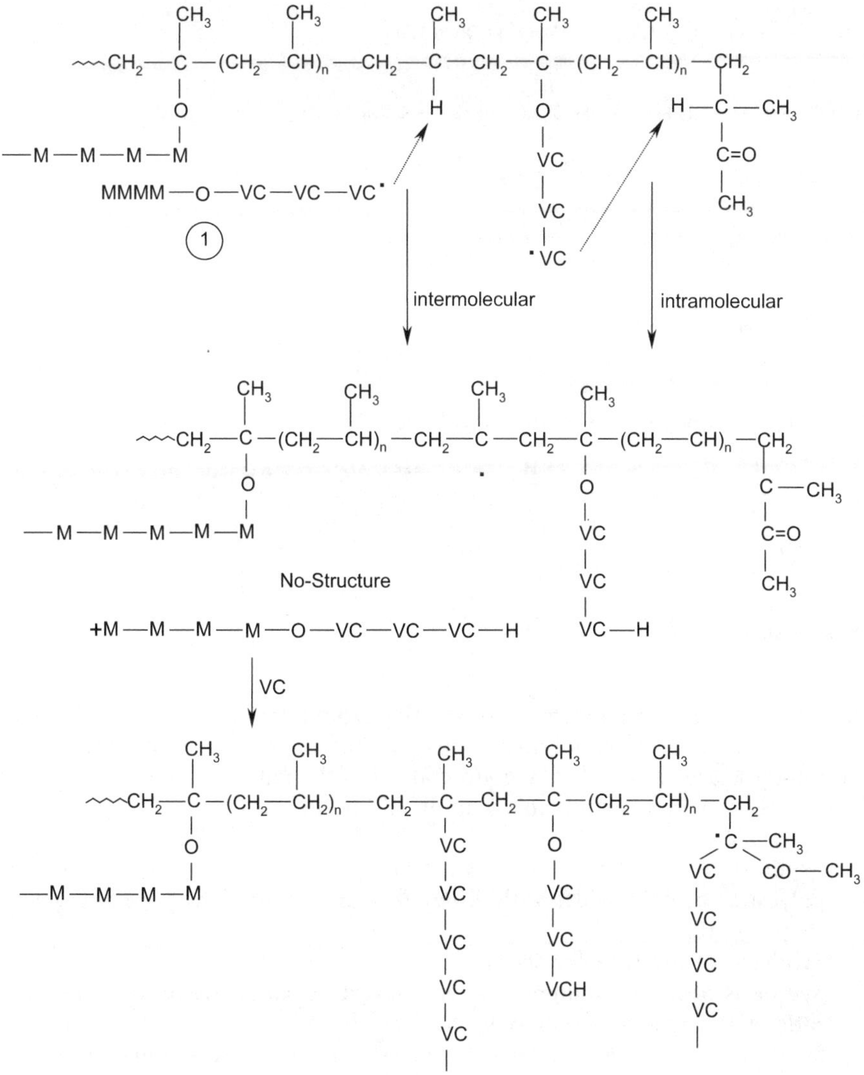

MMMM—O—VC—VC—VC•

(1)

intermolecular

intramolecular

No-Structure

+M—M—M—M—O—VC—VC—VC—H

VC

Scheme 39

Firstly, Aharoni et al. [128] described a process of ozonization of polymers or copolymers containing unsaturations in a mixture of two solvents. One of them is inert to ozone and the other one is less reactive than polymer double bonds but more reactive than the single C-C bonds of this polymer. A typical solvent mixture is composed of toluene and 1,1,2,2,-tetrachloroethane, or xylene and decaline. This mixture permits one to control the attack of polymer only onto unsaturations and not to produce unstable sites in the polymer backbone (as peroxides or hydroperoxides coming from single bond C-C attack) which could decompose in a second step producing undesirable by-products. In this way, only unsaturations are reacted and

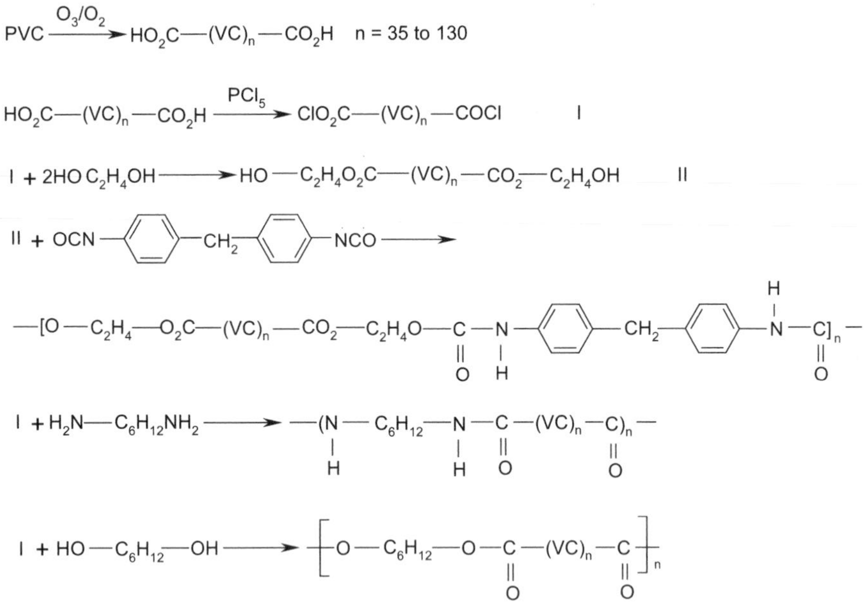

$$PVC \xrightarrow{O_3/O_2} HO_2C—(VC)_n—CO_2H \quad n = 35 \text{ to } 130$$

$$HO_2C—(VC)_n—CO_2H \xrightarrow{PCl_5} ClO_2C—(VC)_n—COCl \qquad I$$

$$I + 2HO\,C_2H_4OH \longrightarrow HO—C_2H_4O_2C—(VC)_n—CO_2—C_2H_4OH \qquad II$$

$$II + OCN—C_6H_4—CH_2—C_6H_4—NCO \longrightarrow$$

$$—[O—C_2H_4—O_2C—(VC)_n—CO_2—C_2H_4O—C(=O)—NH—C_6H_4—CH_2—C_6H_4—NH—C(=O)]_n—$$

$$I + H_2N—C_6H_{12}NH_2 \longrightarrow —(NH—C_6H_{12}—NH—C(=O)—(VC)_n—C(=O))_n—$$

$$I + HO—C_6H_{12}—OH \longrightarrow \left[O—C_6H_{12}—O—C(=O)—(VC)_n—C(=O)\right]_n$$

Scheme 40

lead finally to α-ω oligomers when starting from linear polymers. In the particular case of graft copolymers, mixtures of mono and multi functional compounds are obtained. The good control of the nature of the obtained extremities (di acids) permits many applications such as:

- Synthesis of telechelic oligomers usable in the synthesis of multiblock polyesters or polyamides with different segments presenting specific physical properties
- Synthesis of additives for epoxy resins
- Synthesis of macromonomers when converting acid extremities in unsaturations of vinylic or acrylic type
- Synthesis of macrodiols after selective reduction of acid terminal groups in hydroxyl groups
- Synthesis of ionomer resins after neutralization of acid groups by metallic cations

Other authors have also been interested in this field of research and we will sum up the main results.

The processes described in Table 2 present a peculiar interest in the working out of new materials as polyurethanes. These last polymers are very often based on macro diols coming from polyethers or polyesters, α-ω functional polyolefins being relatively uncommon. Hence, Rhein and Ingham [139] prepared macrodiols by ozonization of polyisobutylene in CCl_4 at

Table 2 Summary of studies focused on ozonization of polydienes

Polymer	Experimental Conditions	Post Ozonization Treatment	Obtained Products	Ref.
Poly 1-4 isoprene	Inert solvent −70 °C to 30 °C	Reduction with $LiAlH_4$	α,ω-Macrodiols	[129, 130]
1-4 *cis*-Polybutadiene	Aromatic solvent 10 °C	Reduction with $(CH_3OCH_2CH_2O)_2AlNaH_2$	α,ω-Macrodiols	[131]
Acrylonitrile-butadiene copolymer	Tetrahydrofuran 15 °C	Reduction with Na borohydrides	α,ω-Macrodiols for polyurethanes	[132–133]
Polyisobutylene	Cyclohexane	Reduction with H_2/Ni Raney	α,ω-Macrodiols	[134]
Polydienes	Alkanes, −20 °C	$LiAlH_4$ or Ni/H_2	α,ω-Diol oligomers	[135]
Polyisobutylene	Suspension in hexane	-	α,ω-Acid oligomers	[136]
Polybutene	Suspension in hexane	Thermal decomposition of peroxides and reduction with Ni Raney or $LiAlH_4$	Polyols	[137, 138]

25 °C followed by reduction using a hydride like $(CH_3OCH_2CH_2)_2AlNaH_2$. The diols prepared according to this technique present an average molecular weight of 2000 g mol^{-1}, starting from a polymer of 82,000 g mol^{-1}. After reaction in bulk with tolylene diisocyanate, these diols give good binders for solid propellants. The same reaction procedure has been applied to copolymers of ethylene-propylene-butadiene and permitted to obtain diols with average molecular weights around 8000 g/mol usable in polyurethane formulation [140, 141]. The macrodiols listed in Table 2 are obtained according to different types of reduction but the final applications are mainly the polyurethanes [142].

More recently, Weider et al. [143] prepared polycarbonates starting from chlorinated rubbers. The diacids obtained after ozonization of these polymers are reacted with phosgene and Bisphenol A to give a polycarbonate used as additive in commercial polycarbonate or in blends of polycarbonate with polyisobutylene.

The chemical modification of polymers with a view to confer them new properties presents a great interest. In this area, the activation of natural or synthetic polymers by ozone followed in a second step by grafting of monomers or reactive molecules in radical medium constitutes a field of important developments, mainly in terms of patents. We will give here the main outlines.

4.2 Synthesis of Graft Copolymers. Modification of Polymers

Janssen et al. [144] focused their work on ozonization of polyvinyl lactam, grafting with hydrophilic methacrylic monomers for applications in the field of contact lenses and other products used in the medical domain. The most studied polymer remains the poly-*N*-vinyl pyrrolidone which is ozonized either in solid state or in aqueous solution. This activation step leads to three hydroperoxides per chain but also to chain scissions. The resulting product is formulated with different mixtures of methacrylic and dimethacrylic monomers to graft them onto activated polymer by UV initiation. Using dimethacrylic monomers lead to perfect cross-linked polymers presenting excellent resistance to solvents. Unfortunately, the mechanisms of action of ozone onto polyvinyl lactams do not seem to have been studied in detail.

In a similar way, Iwasaki et al. [145] modified the surface of polyamide 6 films or fibers after ozonization. These authors grafted in a second step hydrophilic monomers such as acrylamide, and also vinyl acetate or methyl methacrylate. Once more, oxidation mechanisms and preferential sites of attack of ozone have not been clearly described in the literature.

The Idemitsu Kosan Co. Ltd developed an original process of grafting of monomers (for example styrene) onto aromatic polyesters like poly bisphenol A terephthalate. In this process, the time of treatment of the cited polymer remains long what is due to its aromatic character well known to be more resistant to ozone than aliphatic polymers [146]. It can be noted that the graft copolymers obtained according to this procedure present excellent mechanical properties and good transparency.

The field of research of improving weatherability of polymers containing organic dyes has been subjected to numerous studies. One of the solutions proposed by Asahi Chemical Industry and Co. consists, in a first step, in the ozone activation of polymers such as polyolefins, polyvinyl chloride, aromatic polyesters, and in a second step, in grafting of organic dyes presenting unsaturations (acrylic, methacrylic, or vinylic). Such graftings prevent migration of dyes in the polymer and improve their stability and aging [147].

Only little work has been focused on the action of ozone onto non-hydrocarbon polymers. This area of research has been investigated by Karandinos et al. [148] who cross-linked polydimethylsiloxanes (PDMS) bearing silane groups Si-H after ozonization. These workers studied the reaction shown in Scheme 41.

Starting from this concept, the authors developed different types of cross-linked elastomers starting from PDMS bearing Si-H bonds either located on the polymer backbone or at the extremities. The PDMS have been cross-linked using phenyl silane (Scheme 42).

The phenyl silane reacts with ozone to give a tri silanol ØSi$(OH)_3$ which

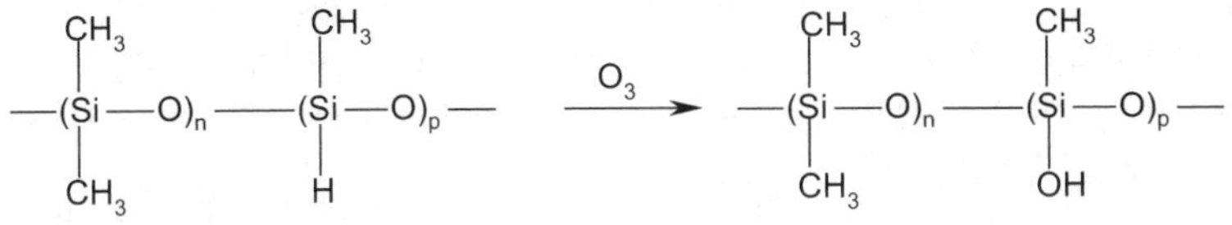

$$-(\underset{CH_3}{\overset{CH_3}{Si}}-O)_n-(\underset{H}{\overset{CH_3}{Si}}-O)_p- \xrightarrow{O_3} -(\underset{CH_3}{\overset{CH_3}{Si}}-O)_n-(\underset{OH}{\overset{CH_3}{Si}}-O)_p-$$

Scheme 41

condenses with PDMS bearing silanols groups coming from the action of ozone onto PDMS including Si-H groups. The authors increased the cross-linking rate of these elastomers when adding some agents such as $C_6H_5POCl_2$ or P_2O_5 which provide the total conversion of the last Si-OH groups [149]. Materials obtained according to this procedure present outstanding thermal resistance while keeping low T_g.

In terms of industrial developments, the reactivity of PDMS with ozone has been turned to account for the elaboration of contact lenses or other similar products used in medical field. The main difficulty in this application is the biocompatibility of polymers with the human body. This lack of synthetic polymers has been overcome by grafting hydrophilic monomers on the surface of polymers, grafting in the depth having to be absolutely avoided. Hence, Bertrand et al. [150] grafted *N*-vinyl pyrrolidone onto ozonized polydimethylsiloxane. This monomer being incompatible with the polymer, its penetration in the depth of the polymer remains very low and grafting is mainly observed at the surface of the material. Moreover, in order to avoid the formation of homopolymers coming from the initiation of radicals OH˙ resulting from the decomposition of hydroperoxides, the authors

where phenyl rings decompose slowly in oxalic acid HO-C(=O)-C(=O)-OH

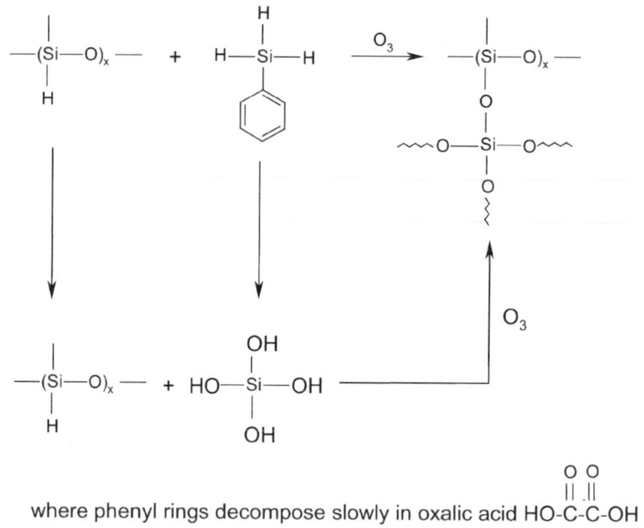

Scheme 42

worked in basic aqueous medium containing additives such as ferrous ammonium sulfate. This system is a well known composition used to limit the formation of homopolymer by a Redox mechanism converting ferrous ions in ferric ones (Scheme 43).

The effect of this composition has been subsequently improved to regenerate the oxidized metal in its reduced form by formation of a complex with squaric acid [151, 152] (Scheme 44).

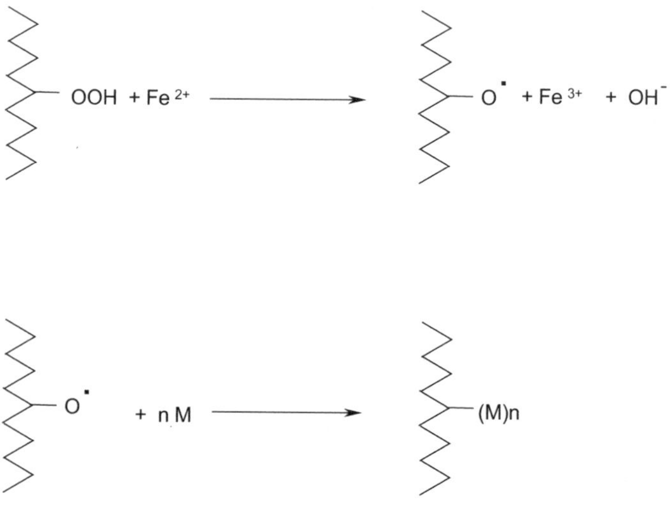

$$\text{~OOH} + Fe^{2+} \longrightarrow \text{~O}^{\bullet} + Fe^{3+} + OH^{-}$$

$$\text{~O}^{\bullet} + n\,M \longrightarrow \text{~(M)n}$$

Scheme 43

This last compound also prevent the penetration of metal salts in the depth of the polymer. Otherwise, other authors such as Janssen et al. [153, 154] investigated the chemical modification of polydimethylsiloxane to prepare materials used in the medical field. Hence, they modified the wettability of PDMS by a treatment of surfaces with ozone followed by UV decomposition of the formed hydroperoxides in the presence of transfer agents such as alcohol or mercaptans.

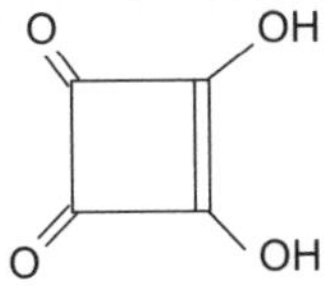

Squaric Acid $C_4H_2O_4$

$$C_4H_2O_4 + 8\ H_2O + 10\ Fe^{3+} \longrightarrow 4\ H_2CO_3 + 10\ Fe^{2+} + 10\ H^+$$

Scheme 44

The authors demonstrated also the conversion of peroxides into OH functions which greatly improve the wettability of materials. It is interesting to note that the activation of the polymer by ozone is operated in fluorinated solvents (Freon 113) in which the polymer is not soluble but ozone is very soluble.

The same authors developed a process of encapsulation of polymers swelled by halogenated solvents in which ozone is greatly soluble but not monomers to be grafted. After ozonization of polymers swelled in solvents, mixtures of mono unsaturated or di unsaturated monomers are added to the activated polymers. Then, grafting is operated by UV irradiation. Grafting is mainly located at the surface of the starting polymer what prevents the modification of its intrinsic properties. This process permits to produce hydrophilic polysiloxanes used in medical applications (contact lenses, tubes, catheters, etc.).

An important part of the works focused on the synthesis of graft copolymers starting from ozone activated polymers is devoted to polyolefins and chlorinated polymers. These graft copolymers have been applied in different areas such as compatibilizers for polymer blends, hot melt adhesives, ion-exchange membranes. In all the described processes, the same general procedure consists in the ozonization of a polymer backbone in solid state, in solution or in suspension and in a second step, in the grafting of one or several monomers. This last step can be achieved in the molten state or in solution. The following table gives examples of different graft copolymers prepared according to this conventional procedure (Table 3).

The activation of polyolefins has been extensively studied by numerous authors as Boutevin and Robin. As mentioned in Table 3, Robin [174] devel-

Table 3 Summary of studies devoted to graft copolymers from ozonized polyolefins or chlorinated polymers

Polymer	Experimental ozonization conditions	Grafted monomer	Experimental grafting conditions	Application	Ref.
EVA crosslinking	Suspension	Styrene	Solution	Emulsifier	[155]
EVA crosslinking	Suspension	AA	Solution	Adhesive	[156]
PVC	Solid state	Styrene	Solution	Emulsifier	[157]
PE	Solid state	Styrene	Bulk	Emulsifier	[158–160]
PE	Solid state	Methyl acrylate	Bulk	Emulsifier	[161, 162]
PE	Solid state	AA MAGLY VAC HEA	Bulk	Adhesive	[163, 164]
PVDF	Solid state	AA or MAGLY or MMA	Bulk	Emulsifier	[165, 166]
PE	Solid state	AA or MADAME	Bulk or suspension	Ion exchange membrane	[167]
PE	Solid state	Vinyl chloride	Suspension		[168]
PE	Solid state	Fluorinated monomers	Suspension		[169]
PE or PP	Suspension	Acrylamide or AA	Suspension		[170, 171]
E.P.D.M.	Solution	Methyl methacrylate	Solution	Transparent elastic sheet	[172]
E.P.	Suspension	Styrene	Suspension	Improving of wettability	[173]

AA: acrylic acid
HEA: hydroxy ethyl acrylate
MAGLY: glycidyl methacrylate
MADAME: dimethyl amino ethyl methacrylate
HEMA: hydroxy ethyl methacrylate

oped original work using the principle of the living radical polymerization to ozonized polymers. Hence, starting from activated polyethylene, they grafted monomers such as styrene in the presence of nitroxyl radicals, these last compounds permitting a real control of the copolymerization kinetics and of the chain length. Consequently, the grafting rates are superior to those obtained with conventional experimental conditions and the living character has been demonstrated. So, copolymers obtained after copolymerization and further purifications are able to initiate a new polymerization when a monomer is added to the copolymer and heated. This ability to start the polymerization again when a supplementary quantity of monomer is added proof of the living character of this type of copolymerization (Scheme 45).

Some authors took advantage of the reactivity of polymers with ozone to recycle some peculiar plastic wastes coming from different origins. In this way, Fargère [155, 156] succeeded in degradation of cross-linked ethylene-vinyl acetate copolymers used in the production of sport shoes. This operation was achieved on grinded industrial wastes in suspension in CCl_4. In a second

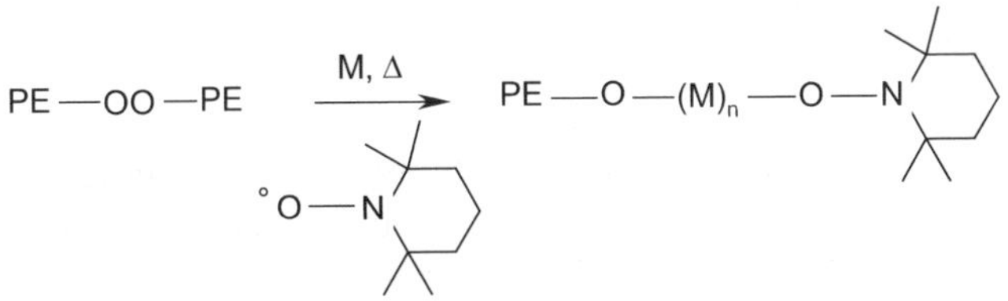

PE—OO—PE $\xrightarrow{M,\ \Delta}$ PE—O—$(M)_n$—O—N

°O—N

Scheme 45

step, they grafted on the activated copolymers different monomers such as styrene or acrylic acid. The resulting products can be used in aqueous emulsion usable in the field of adhesives or coatings. So, ozone constitutes an original way to degrade this type of cross-linked polymers and to graft them with monomers that confer new properties to them.

In the same manner, Platz [175] investigated the use of ozone to depolymerize different vulcanizated rubbers (wastes coming from tires for example). The first step consists in the grinding of tires which are ozonized. In a second step, these granulates are heated under reduced pressure with catalysts like $MgCl_2$ or $AlCl_3$ to produce a mixture of monomers. This depolymerization reaction is conducted at temperatures lower than that applied during conventional pyrolysis. The use of a catalyst and of moderated temperatures permits to obtain products which compositions are less complex than those obtained with more drastic conditions. This process can also be applied to cross-linked polyethylene or to polyurethane foams to convert them in α-ω macrodiols. This type of treatment can be realized in solvent such as acetone but in this case, the resulting products are more complex since solvent is sensitive to ozone and can be degraded. However, solvent is necessary to make soluble macrodiols as soon as they are produced in the medium [176].

Boutevin et al. [177–180] treated different types of wastes of polyolefins (more often low density polyethylene) with a mixture air/ozone. They focused their studies on the quantification of the formed oxygenated species based on colorimetric titration using stable radicals such as diphenylpicrylhydrazyl. They investigated the influence of mineral compounds (iron oxide, for example) used as catalysts for oxidative reactions. The ozonized polymers have been used as binders for composites materials containing mineral materials (sand, stones, etc.).

At last, we must mention studies developed by Robin [181–183] who added ozonized polyolefins to bituminous materials to obtain new binders for road construction. They demonstrated that peroxides synthesized during the ozonization step decompose and react with specific compounds contained in bitumen. It follows that it can be observed a very sharp improvement in mechanical properties of binders and also of mechanical behavior and aging of composite materials used in road construction.

5 Modification of Natural Polymers Using Ozone

The reactivity of compounds like natural oils has been extensively studied with the aim to produce surfactants or resins [184, 185]. The modification of natural compounds, such as natural polymers, by synthetic monomers to confer them new properties has been extensively studied by researchers and presents a great interest. Indeed, natural polymers constitute a renewable resource providing interesting specific properties but affording some lacks, notably in terms of mechanical properties. Hence, researchers tried, either to mix them with synthetic polymers or to modify them by copolymerization. Ozone constitutes a interesting reactant to activate natural compounds like cellulose in order to graft monomers in a second step. However, ozonization of natural polymers remains a very complex chemical reaction and researchers tried recently to elucidate mechanisms using model compounds like methyl pyranosides [186]. We will try to show now the main results in this field when using ozone as activator of polymers.

The main grafting described in the literature and experimental conditions applied can be summarized in Table 4.

In all cases, the authors propose technical solutions permitting one to reach non-negligible grafting rates knowing that the majority of natural starting compounds contain substances such as phenols inhibiting all polymerization of monomers when using conventional methods. So, ozonization

Table 4 Summary of studies focused on the chemical modification of natural polymers using ozone as activator

Material support	Grafted monomers	Comments	Ref.
Cotton	Methyl methacrylate		[187]
Cellulose	Acrylamide	Application in paper making industry. Improvement of fire resistance	[188, 189]
Ligno sulfonates	Styrene, acrylonitrile	Grafting of styrene in methanolic solution	[190, 191]
Gelatine	Styrene		[192]
Cellulose	(Meth)acrylic monomers	Great improvement in mechanical properties and lowering of hygroscopy	[193]
Collagene	Methyl methacrylate		[194]
Lignin	Styrene		[195]
Lignin	Styrene	Ozonization avoids inhibiting effect of lignin on styrene polymerization	[196]

of these natural polymers constitutes an original and interesting way for modification of natural polymers. Moreover, ozonization in methanolic medium or in dimethyl acetamide leads to a sharp increase in grafting rate, what can be explained by the swelling of substrates in solvents.

Yoshida et al. [197] developed thermostable materials based on ozonized lignin. Hence, the authors ozonized lignin to prepare muconic acid or its hemi-ester, as shown in Scheme 46.

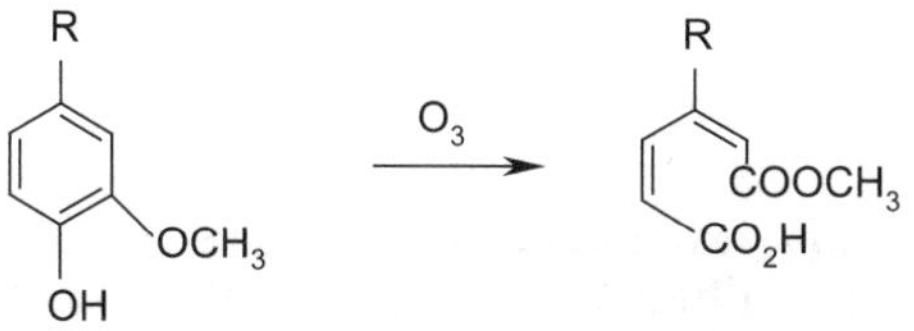

R OH OCH_3 O_3 R $COOCH_3$ CO_2H

Scheme 46

In a second step, a Diels-Alder reaction leads to a maleimide compound (Scheme 47).

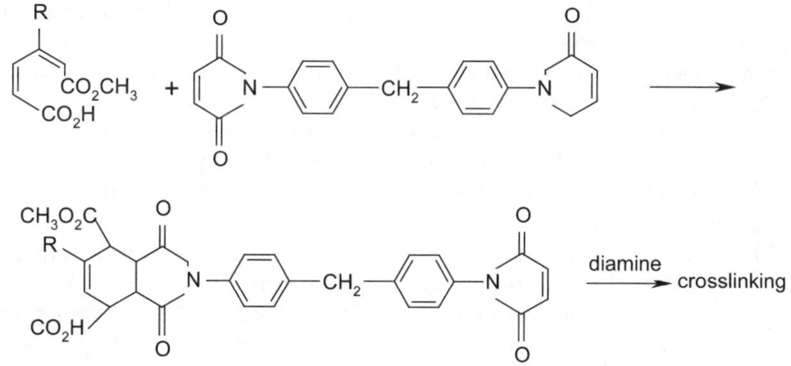

R CO_2CH_3 CO_2H + O O N CH_2 N O

CH_3O_2C R CO_2H O O N CH_2 N O O diamine crosslinking

Scheme 47

Afterwards, the authors prepared also polyamides according to the reaction shown in Scheme 48.

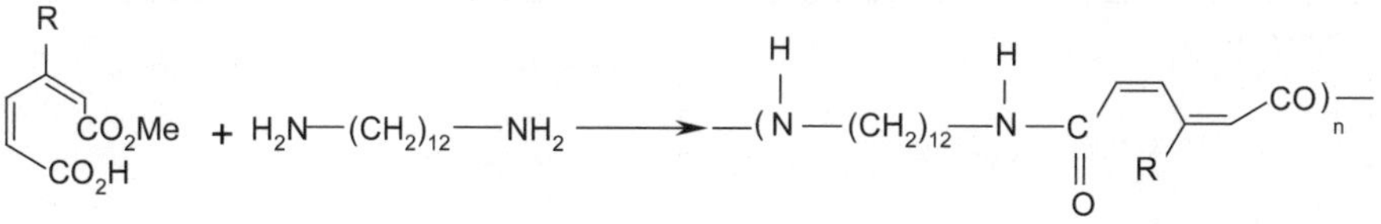

R CO_2Me CO_2H + $H_2N-(CH_2)_{12}-NH_2$ $\longrightarrow$ $-(N-(CH_2)_{12}-N-C$ H H O R $CO)_n-$

Scheme 48

All these syntheses produce interesting products in terms of thermal stability. It must be noted that this muconic acid has been used by the same authors [198] to participate to the hardening of epoxy resins where it is soluble above 120 °C. These resins are mainly used as adhesives for wood.

The treatment of lignin by ozone, and so its functionalization by acid groups provided them outlets, knowing that lignin constitutes a voluminous by-product resulting from wood delignification in the paper industry [199–202].

6 Modification and Treatment of Surfaces by Ozone

The low surface tension of the majority of polymers makes them difficult to paint, to assembly with adhesives, or to print with inks. Table 5 gathers surface tension values of usual commodity plastics [203].

The modification of the chemical composition of polymer surfaces, and thus their wettability with chemical substances, can be realized in different ways: electric discharges more commonly called Corona effect, oxidation by a flame, plasma treatment, UV irradiation and also UV irradiation under ozone atmosphere. Numerous studies have been devoted to the effects of these different treatments. More recently, Strobel et al. [204] compared the effects of these treatments on polypropylene and polyethylene terephthalate using analytical methods such as E.S.C.A., F.T.I.R., and contact angle measurements. They demonstrated that a flame oxidizes polymers only superficially (2–3 nm) whereas treatment realized by plasma effect or Corona effect permits one to work deeply in the polymer (10 nm). The combination of UV irradiation with ozone flux modifies the chemical composition of the polymers to a depth much greater than 10 nm, introducing oxygenated functions into the core of the polymer.

Kulik et al. [205] focused their studies on the identification of chemical species formed during the treatment of polyolefins such as polyethylene or polypropylene by gaseous ozone or ozone in aqueous medium. Experimental conditions have a great influence on the nature of the obtained species. For example, peroxidic functions, carboxylic acids, and ketones have been identified, aldehydes being absent of the surface of the materials. It must be noted here the instability of the peroxidic species formed during the treatment

Table 5 Surface tensions of usual polymers

Polymers	Surface tensions (in mN/m)
Nylon 6–6	46
PET	43
PCTFE	31
PE	31
PTFE	18.5
PHFP	16.2

and their slow degradation in carbonyl groups (ketones, acids). In a same manner, Yu and Song [206] studied the same polymers and demonstrated that the surface oxidation of polyolefins comes from a preferential attack of the tertiary carbon, mainly in the amorphous zones of the material, the acid functions being the major functions at the end of the ozone treatment.

These studies devoted to the reactivity of polymers in relation to their structure have been used by industrials and many patents claimed the use of ozone to improve the welding of polymers or the paintability of polymer surfaces. Mitsubishi Petrochemical Co. [207] developed a process to obtain low density polyethylene based on multilayers where adhesiveness of the different layers is favored by an ozone treatment. Moreover, the process can be improved when using a combination of ozone with reactants such as ammonia, hydrazine which improve in a better way the surface modifications. Thus, the immersion of activated polyolefins in aqueous solution of ammonia confers to surfaces a better hydrophilicity but also permits a very good adhesion of epoxy resins or all other coating [208]. The same company invented a process to oxidize polyolefins with ozone immediately after extrusion and calandring what greatly improves the laminating on other substrates [209]. In another way, Ko et al. [210] described the preparation of surface-modified polymer culture dishes using ion beam irradiation to activate polymer surfaces followed by reaction with ozone to produce hydrophilic functions.

Other industries developed in the same manner the use of ozone for the treatment of surfaces. Hoechst AG claimed a process of ozonization of poly(ethylene-*co*-5-vinyl-2 norbornene) to give polymers usable as protective coatings for metals or plastic materials [211].

7 Conclusion

The aim of this chapter was to show that ozone constitutes a good reagent since the majority of polymers are sensitive to its attack. Polymers like polyolefins react in giving functions such as acid or hydroxyl groups, and also ketones and lactones. Polymers containing unsaturations are the most sensitive and their treatment leads rapidly to ozonides of low stability and involve a rapid decrease in molecular weights of polymers, producing α-ω oligomers. Some authors modified other kinds of polymers like halogenated polymers or polysiloxanes which present a significant reactivity to ozone.

Most authors use ozone with the target to obtain different types of well defined structures starting from natural or synthetic polymers. Most of them ozonize polydienes or polymers containing some unsaturations along the chain, producing chain scissions with chain end reactive groups. The latter permit copolymerization of different monomers and the production of block copolymers. Graft copolymers are most often produced by ozonization of saturated polymers or natural polymer and grafting of monomers after decomposition of the formed peroxides and hydroperoxides. Some authors succeeded in a good control of the chain length and kinetic polymerization with the use of nitroxyl radicals.

Another field of research concerns the modification of polymer surfaces or natural polymers bearing some lacks in properties like dyeing, hydrophilicity, etc. The treatment of such materials by ozone permits to confer them better performance.

Ozone being unstable in temperature, its elimination after reaction can be easily achieved by heating, producing oxygen. Furthermore, taking into account the straightforwardness of its use since polymers can be activated in the solid state, without solvent or purifications, many applications have been developed starting from this reagent.

References

1. Zhang XZ, van Heiningen A (2000) J Pulp Paper Sci 26:335
2. Fournier G (1983) Cinétique de l'oxygène en milieu plasma. Réactivité dans les plasmas. Application aux lasers et aux traitements de surface. Les Editeurs de Physique, Ecole d'Eté d'Aussois, p 297
3. Guenther KF, Sosnovsky G, Brunier R (1964) Anal Chem 36:2508
4. Smits M, Hoefman D (1972) Anal Chem 44:1688
5. Criegee R, Wenner G (1949) Justus Liebigs Ann Chem 564:9
6. Criegee R (1953) Justus Liebigs Ann Chem 583:1
7. Bailey PS (1958) Chem Rev 58:925
8. Criegee R (1951) Abstracts. 120th National Meeting of the American Chemical Society, New York, p 22M
9. Criegee R, Kerckow A, Zinke H (1955) Chem Ber 88:1878
10. Criegee R (1957) Rec Chem Prog 18:111
11. Criegee R (1962) Peroxide reaction mechanisms. Edwards JO (ed). Wiley (Interscience), New York, pp 29–39
12. Pryde EH, Cowan JC (1971) Top Lipid Chem 2:98
13. BaileyB PS (1978) Ozonization in organic chemistry. Academic Press, New York, vol I, p 45
14. Young WG, McKinnis AC, Webb ID, Roberts JD (1946) J Am Chem Soc 68:293
15. Murray RW, Youssefyeh RD, Story PR (1967) J Am Chem Soc 89:2429
16. Bailey PS, Ward JW, Carter TP Jr, Nieh E, Fisher CM, Khashab AY (1974) J Am Chem Soc 96:6136
17. Bauld NL, Thompson JA, Hudson CE, Bailey PS (1968) J Am Chem Soc 90:1822
18. Tits M, Bruylants A (1948) Bull Soc Chim Belg 57:50
19. Goodwin SM, Johnson NM, Witkop B (1953) J Am Chem Soc 75:4273
20. Kolsaker P (1965) Acta Chem Scand 19:223
21. Lorenz O, Parks CR (1965) J Org Chem 30:1976
22. Fliszar S, Carles J (1969) Can J Chem 47:3921
23. Fliszar S, Carles J (1969) J Am Chem Soc 91:2637
24. Murray RW, Williams GJ (1968) Adv Chem Ser 77:32
25. Greenwood FL (1966) J Am Chem Soc 88:3146
26. Bailey PS, Rustaiyan A, Ferrell TM (1976) J Am Chem Soc 98:638
27. Bailey PS (1982) Ozonization in organic chemistry. Academic Press, New York, vol II, p 371
28. Griesbaum K, Keul H (1975) Angew Chem Int Ed Engl 14:716
29. Griesbaum K, Hofmann P (1976) J Am Chem Soc 98:2877
30. Griesbaum K, Keul H (1975) Angew Chem 87:748
31. Kwart H, Hoffman DM (1966) J Org Chem 31:419

32. Criegee R, Günther P (1963) Chem Ber 96:1564
33. Bailey PS, Lane AG (1967) J Am Chem Soc 89:4473
34. Enzell CR, Thomas BR (1965) Tetrahedron Lett 225
35. Enzell CR, Thomas BR (1965) Acta Chem Scand 19:1875
36. Toby FS, Toby S (1974) Int J Chem Kinet 6:417
37. Toby FS, Toby S (1975) Int J Chem Kinet Symp 1:197
38. Bailey PS (1978) Ozonization in organic chemistry. Academic Press, New York, vol II, p 1
39. Criegee R (1953) Liebigs Ann Chem 583:1
40. Nakagawa TW, Andrews LJ (1960) J Am Soc 82:26,941
41. Razumovskii SD, Niazashvili GA, Tutorskii IA, Yur'ev YN (1971) Vysokomol Soedin Ser A 13:190
42. Sawaki Y, Ogata Y (1978) J Am Chem Soc 100:856
43. Haaijman PW, Wibaut JP (1941) Recl Trav Chim Pays-Bas 60:842
44. Harries CD (1905) Liebigs Ann Chem 343:369
45. Wibaut JP (1950) Bull Soc Chim Fr 996
46. Wibaut JP (1956) J Chim Phys Chim Biol 53:111
47. Wibaut JP (1955) Ind Chim Belge 20:3
48. Wibaut JP (1957) Chimia 11:298
49. Wibaut JP, Sixma FLJ, Kampschmidt LWF, Boer H (1950) Recl Trav Chim Pays-Bas 69:1355
50. Sixma FLJ, Boer H, Wibaut JP (1951) Recl Trav Chim Pays-Bas 70:1005
51. Wibaut JP, Sixma FLJ (1952) Recl Trav Chim Pays-Bas 71:761
52. Bailey PS (1982) Ozonization in organic chemistry. Academic Press, New York, vol II, p 20
53. Nakagawa TW, Andrews LJ, Keefer RM (1960) J Am Chem Soc 82:269
54. Bernatek E, Karlsen E, Ledaal T (1967) Acta Chem Scand 21:1229
55. Whiting MC, Bolt AJN, Parish JH (1968) Adv Chem Ser 77:4
56. Murray RW (1979) Singlet oxygen. Wasserman HH, Murray RW (eds). Academic Press vol 3, pp 93–114
57. Stary FE, Emge DE, Murray RW (1976) J Am Chem Soc 98:1880
58. Stary FE, Emge DE, Murray RW (1974) J Am Chem Soc 96:5671
59. Briner E, Demolis A, Paillard H (1931) Helv Chim Acta 14:794; (1932) Helv Chim Acta 15:201
60. Briner E (1958) Adv Chem Ser 21:184
61. Briner E (1940) Helv Chim Acta 23:590
62. Briner E, Fliszar S (1959) Helv Chim Acta 42:2063
63. Briner E, Fliszar S (1959) Helv Chim Acta 42:2753
64. Erikson RE, Bakalik D, Richards C, Scanlon M, Huddleston G (1966) J Org Chem 31:461
65. White HM, Bailey PS (1965) J Org Chem 30:3037
66. Paillard H, Briner E (1942) Helv Chim Acta 25:1528
67. Taube H (1957) Trans Faraday Soc 53:656
68. Stoll M, Scherrer W (1930) Helv Chim Acta 13:142
69. Hamilton GA, Ribner BS, Hellman TA (1968) Adv Chem Ser 77:15
70. Syroezhko AM, Korothova NP, Vikhorev AA, Proskuryakov VA (1978) Zh Prikl Khim (Leningrad) 51:2562; (1978) J Appl Chem (USSR) 51:2442
71. Komissarov VD, Galimova LG, Denisov ET (1974) Kinet Katal 15:1063; (1974) Kinet Catal 15:944
72. Gerchikov AY, Komissarov VD, Denisov ET, Kochemasova GB (1972) Kinet Katal 13:1126; (1972) Kinet Catal 13:1012
73. Aleksandrov YA, Tarunin BI (1977) Usp Khim 46:1721; (1977) Russ Chem Rev 46:905
74. Spialter L, Pazdernik L, Bernstein S, Swansiger WA, Buell GR, Freeburger ME (1972) Adv Chem Ser 112:65
75. Spialter L, Pazdernik L, Bernstein S, Swansiger WA, Buell GR, Freeburger ME (1971) J Am Chem Soc 93:5682

76. Spialter L, Austin JD (1965) J Am Chem Soc 87:4406
77. Spialter L, Austin JD (1966) Inorg Chem 5:1975
78. Austin JD, Spialter L (1968) Adv Chem Ser 77:26
79. Spialter L, Swansiger WA (1968) J Am Chem Soc 90:2187
80. Buell GR, Spialter L, Austin JD (1968) J Organomet Chem 14:309
81. Spialter L, Swansiger WA, Pazdernik L, Freeburger ME (1971) J Organomet Chem 27:C25
82. Dexheimer EM, Spialter L (1975) J Organomet Chem 102:21
83. Bailey PS (1982) Ozonization in organic chemistry. Academic Press, New York, vol II, p 330
84. Kalkis V, Zicans J, Kalnis M, Ivanova Tatjana (2000) LV Pat 12468 C.A. 2001 135:33986
85. Oueslati R, Catoire B (1991) Eur Polym J 27:331
86. Oueslati R, Bahri H (1992) Eur Polym J 28:1247
87. Carlsson DJ, Wiles DM (1970) J Polym Sci B8:419
88. Razumovski SD, Zaikov GE (1971) Eur Polym J 7:275
89. Michel A, Monnet C (1981) Eur Polym J 17:1145
90. Livanova NM, Popova ES, Ledneva OA, Popov AA (2000) Inter J Polym Mater 47:279
91. Rabek JF, Lucki J, Ranby B (1979) Eur Polym J 15:101
92. Egorova GG (1985) Vest Leningr Univ Fiz Khim 4:66 C.A. 1985 104:150,508
93. Brydson JA (1978) Rubber chemistry. London, Applied Science Publishers
94. Rayer RW, Latiimer RP (1990) Rubber Chem Tech 63:426
95. Cataldo F, Ori O (1995) Polym Deg Stab 48:291
96. Cataldo F (1996) Polym Degrad Stab 53:51
97. Lacoste J, Claudie A, Siampiringue N, Lemarie J (1994) Eur Polym J 30:433
98. Anachkov MP, Rakovski SK, Stefanova RV (2000) Polym Deg Sta 67:355
99. Tyurina YE, Kukovinets OS, Sigaeva NN, Volodina VP, Abdullin MI, Yu B, Prochukhan YA (2000) Bashkrirskii KhimichesKii Zhurnal 7:41 C.A. 2000 133:296,688
100. Stephens WD, Intosh CR, Taylor CO (1968) J Polym Sci Part A-1,6:1037
101. Cataldo F, Ricci G, Crescenzi V (2000) Polym Degrad Stab 67:421
102. Ho KW (1986) J Poly Sci, Part A Polym Chem 2462
103. Anachkov MP, Rakovsky SK, Shopov DM (1985) Polym Deg Stab (10) 25
104. Anachkov MP, Rakovsky SK, Shopov VDM (2000) Polym Deg Stab (67) 355
105. Peters GS, Gorham SD, Smith FJ, Fraser AM, Colclough ME, Millar R (2000) Propellants Explosives Pyrotechnics 25:191 C.A. 2000 134:740,760
106. Razumovskii SD, Zaikov GE (1983) Dev Polym Deg Stab 6:239 C.A. 103:7075
107. Saito T, Niki E, Shiono T, Kamiya Y (1978) Bull Chem Soc Jpn 51:1153
108. Razumovskii SD, Karpukhin ON, Kefeli AA, Pokholok TV, Zaikov GY (1971) Vysokomol Soyed A 13:782
109. Kefely AA, Rakovski SK, Shopov DM, Razumovskii SD, Rakovski RS, Zaikov GE (1981) J Polym Sci Part A Polym Chem 19:2175
110. Peeling J, Jazaar MS, Clark DT (1982) J Polym Sci Part A Polym Chem 20:1797
111. Kefely AA, Rakovski SK, Shopov DM, Razumovskii SD, Zaikov GE (1981) J Polym Sci 19:2175
112. Tanaka Y, Sato H, Nakafutami Y (1981) Polymer 22:1721
113. Michel A, Schmidt G, Castaneda E, Guyot A (1975) D Angew Makromol Chem 47:61
114. Michel A, Castaneda E, Guyot A (1975) J Macromol Sci Chem A12:227
115. Michel A, Schmidt G, Guyot A (1973) ACS Symp Polym Preprints 14:665
116. Landler Y, Lebel P (1960) J Polym Sci 48:177
117. Cheradame H (1989) In: Goethals EJ (ed) Telechelic polymers. CRC Press, Bota Raton, chap 7, p 141
118. Rimmer S, Ebdon JR (1995) Macromol Rep A32:831 (Supplements 5 and 6)
119. Caffetera LFR, Lombardo JD (1994) Int J Chem Kin 24:503
120. Brandrup BJ, Immergut EH (eds) (1989) Polymer handbook, 3rd edn. Wiley, New York, chap II, p 26
121. Rimmer S, Ebdon JR (1996) J Polym Sci Part A Polym Chem 34:3573

122. Rimmer S, Ebdon JR (1996) J Polym Sci Part A Polym Chem 34:3591
123. Rimmer S, Ebdon JR (1997) J Chem Res Synop 11:408
124. Smets G, Weinand G, Deguchi S (1978) J Polym Sci Polym Chem 16:3077
125. Smets G, Weinand G (1978) J Polym Sci Polym Chem 16:3091
126. Michel A, Castaneda E, Guyot A (1979) Eur Polym J 15:935
127. Brevet Français (1973) 2 235 952 PCUK C.A. 1975 82:140,827
128. Aharoni SM, Prevorsek DC, Schmitt GJ (1979) Eur Pat 012316 C.A. 1983 98:186,999
129. Odinokov VN, Kukovinets OS, Zhemaiduk LP (1975) Otkrytiya Izobret Prom Obraztsy Tovarnye Znaki 52:76 C.A. 1975 83:180,777
130. Odinokov VN, Kukovinets OS, Zhemaiduk LP (1978) Otkrytiya Izobret Prom Obraztsy Tovarnye Znaki 55(4):91 C.A.1978 88:122,101
131. Yukuta T, Ohhashi T, Taniguchi Y, Arai K (1974) Bridgestone Tire Co Ltd Jpn Pat 49,034,592 C.A. 1974 81:92,749
132. Kojima H, Fujio R, Yukita T, Oonishi A (1972) Bridgestone Tire Co Ltd Jpn Pat 47,036,273 C.A. 1973 78:4945
133. Tikhomirov BI, Baraban OP, Yakubchik AI (1969) Vysokomol Soedin A11:304 C.A. 1969 70:97,759
134. Dubois DA (2000) Shell International Research PCT WO 2,000,032,645 C.A. 2000 133:18,683
135. Tanaka Y, Sato H, Mita K, Shimizu M (1986) Kuraray Co Ltd Jpn Pat 61,136,507 C.A.1987 106:19,229
136. Weider R, Scholl T, Kohler B, Bayer AG (1996) US 5,484,857 C.A. 1995 123:289,243
137. Godt HC (1967) Monsanto Co.Fr 1,497,289 C.A. 1970 69:28,239
138. Godt HC (1969) Monsanto Co US 3,429,936 C.A. 1969 70:78,733
139. Rhein RA, Ingham JD (1978) California Institute of Technology US 4,118,427 C.A. 1979 90:24,526
140. Greene RN, Sohl E, Dupont de Nemours (1974) US 3,857,826 C.A. 1975 82:99,704
141. Ver Strate G (1976) Exxon Research and Engineering Co US 4,358,566 C.A. 1983 98: 35,184
142. Macias A, Rubio B (1983) Inst Plast Caucho Rev Plast Mod 45(322):412 C.A. 1983 99:71,952
143. Weider R, Koehler B, Ebert W, Scholl T, Klauss H, Bayer AG (1997) Eur Pat 0,751,167 C.A. 1997 126:144,958
144. Janssen RA (1988) Ciba Geigy Corporation US 4,791,175 C.A. 1987 107:218,290
145. Iwasaki T, Tomita K, Ueda Y (1993) Nippon Ester Ltd Kobunshi Ronbunshu 50:115 C.A. 1992 118:214,084
146. Idemitsu (1984) Kosan Co Ltd Jpn Patent 59,100,117 C.A. 1984 101:211,910
147. Asahi Chemical Industry Co Ltd GB Patent 1,097,363 (1968) C.A.1968 68:60,487
148. Karandinos A, Farris RJ, McCarthy TJ (1994) Polym Prepr Am Chem Soc Div Polym Chem 35:695
149. Mc Gregor RR, Warrick L (1949) US Patent 2,459,387 C.A. 1949 43:14,334
150. Bertrand W (1992) WO Patent 92 10,533, C.A. 1993 118:61,671
151. Mayhan KG, Janssen RA, Bertrand WJ (1982) US Patent 4,311,573 C.A. 1982 96:104,969
152. Evtushenko AM, Timofeeva GV, Chikhacheva IP, Stravrova SD, Zuko VP (1991) B33(3) 215 C.A. 1992 115:30,077
153. Janssen RA (1990) Ciba Geigy. Eur Pat 378,513, C.A. 1991 114:88,744
154. Janssen RA (1990) Ciba Geigy. Eur Pat 378,512, C.A. 1991 114:88,743
155. Fargère T, Abdennadher M, Delmas M, Boutevin B (1995) Eur Polym J 31:923
156. Fargère T, Abdennadher M, Delmas M, Boutevin B (1995) Eur Polym J 31:489
157. Boutevin B, Pietrasanta Y, Robin JJ, Pollet T (1988) Eur Polym J 24:953
158. Sarraf T (1988) NORSOLOR, FR 2 640 986 C.A. 1994 114:44,329
159. Boutevin B, Pietrasanta Y, Robin JJ, Pabiot J, Leseur B (1992) Makromol Chem Makromol Symp 57:371
160. Boutevin B, Pietrasanta Y, Sarraf T (1987) Angew Makromol Chem 148:195
161. Boutevin B, Pietrasanta Y, Sarraf T (1988) Angew Makromol Chem 162 175

162. Boutevin B, Pietrasanta Y, Taha M, Sarraf T (1985) Polym Bull 14:14
163. Boutevin B, Robin JJ (1990) Eur Polym J 26:559
164. Bodart V, Oreins JM, Laurent G, Declerck F, Malinova A, Torres N, Boutevin B (2000) SOLVAY, WO 2,002,051,633 C.A. 2002 137:104,056
165. Boutevin B, Robin JJ, Serdani A (1992) Eur Polym J 28:1507
166. Boutevin B, Pietrasanta Y, Robin JJ, Serdani A (1991) Eur Polym J 27:815
167. Elmidaoui A, Boutevin B, Belcadi S, Gavach C (1991) J Polym Sci Part A Polym Chem B29:705
168. Boutevin B, Pietrasanta Y, Taha M, Sarraf T (1983) Makromol Chem 184:2401
169. Boutevin B, Mouanda J, Pietrasanta Y, Taha M (1985) Eur Polym J 21:181
170. Dasgupta SJ (1990) Appl Polym Sci 41(1/2):233
171. Wang Y, Kim JH, Choo KH, Lee YS (2000) J Membrane Sci 169:269
172. Masaaki T (1972) SUMITOMO Chemical Co. Ltd, Japan Kokai, 4 807 45 90 C.A.1974 80:60,805
173. Idemitsu Kosan Co. Ltd, (1983) Japan Kokai, 6 012 0712 C.A. 1985 104:149,680
174. Bertin D, Boutevin B, Robin JJ (1998) ATOCHEM, EP 402420 C.A.1999 130:297,115
175. Platz GA (1993) S-P Reclamation, Inc., U.S. 5,369,215 C.A. 1995 122:293,150
176. Popova IA, Galstan GA, Popov AF (1982) Izv Vyssh Uchebn Zaved, Khim Tekhnol 25:1520 C.A.1983 98:90,343
177. Boutevin B, Hervaud Y, Lafont J, Pietrasanta Y (1984) Eur Polym J 20:867
178. Boutevin B, Pietrasanta Y, Taha M, Sarraf T (1984) Eur Polym J 20:875
179. Boutevin B, Falgayrettes G, Hervaud Y, Pietrasanta Y (1984) Eur Polym J 20:1067
180. Antoine R, Boutevin B, Hervaud Y, Lafont J, Pietrasanta Y (1984) Eur Polym J 20:1073
181. Boutevin B, Pietrasanta Y, Robin JJ (1987) Eur Polym J 23:525
182. Boutevin B, Pietrasanta Y, Robin JJ (1987) Eur Polym J 23:415
183. Boutevin B, Maliszewicz M, Pietrasanta Y, Robin JJ (1985) SCREG, Eur Pat 4,016,672 C.A. 1986 105:44,925
184. Takeda T (2000) Osaka Municipal Technical Research Institut, Nippon Yukagakkaishi 49:1369 C.A. 2000 134:87,926
185. Khan M, Tomkinson J, Fitchett C, Colin S, Black MJ (2000) Dupont de Nemours, WO 20,000,786,999, C.A. 2000 134:72,367
186. Olkkonen C, Tylli H, Forskahl I, Fuhrmann A, Hausalo T, Tamminen T, Hortling B, Janson H (2000) Hlzforschung 54:397 C.A.2000 133:311,038
187. Yamaoka A, Shigeoka E (1978) Himeji Kogyo Daigaku Kenkyu Hokoku 31:56 C.A. 1979 91:41,076
188. Kokta B, Araneda L, Daneault C (1984) Polym Eng Sci 24:950
189. Oprea S, Dimitriu S, Bulacovschi V (1979) Cellul Chem Technol 13:3
190. Chen R, Kokta BV, Valade JL (1979) J Appl Polym Sci 24:1609
191. Chen R, Kokta BV, Valade JL (1980) J Appl Polym Sci 25:2211
192. Kudaba J, Ciziunaite E, Golsteinaite Z (1967) Liet TSR Aukst Mokyklu Mokslo Darb Chem Technologija 8:163 C.A. 1973 78:85,960
193. Simionescu C, Oprea S (1972) J Polym Sci Part C 37:251
194. Kudaba J, Ciziunaite E, Lukaitis I (1971) Polim Mater IKH Issled Mater Respub Nanc. Tekh Konf 12th 134. Machyulis A (ed) Publisher: Machyulis. Publisher: Kaunas C.A. 1973 78:85,960
195. Katuscak S, Mahdalik M, Hrivik A, Minarik V (1973) J Appl Polym Sci 17:1919
196. Katuscak S, Mahdalik M, Hrivik A, Minarik V (1972) Pap Puu 54:861–870 C.A. 1973 78:99,317
197. Yoshida Y, Kajiyama M, Tomita B, Hosoya S (1990) Mokuzai Gakkaishi 36:440 C.A. 1990 113:154,548
198. Tomita B, Kurozumi K, Takeruma A, Hosoya S (1989) ACS Symp Ser 39:397 (lignin) 496
199. Lee HJ, Tomita B, Hosoya S (1991) Mokuzai Kogyo 46:412 C.A. 1992 116:107,702
200. Matsumoto Y, Minami K, Ishizu A, Mokuzai Gakkaishi (1993) 39:734 C.A. 1994 121:12,014

201. Kurihara K, Nishama N, Wada T (1994) Nishimatsu Constr Co Ltd Jpn Pat 06,063,911 C.A. 1994 121:60,070
202. Kurihara K, Furukawa S, Shirota S (1994) Nishimatsu Constr Co Ltd Jpn 06,155,418 C.A. 1994 121:258,304
203. Zisman WA (1964) Adv Chem Series 43:1
204. Strobel M, Walzak MJ, Hill JM, Iin A, Karbashewski A (1995) CS Lyons J Adhesion Sci Technol 9:365
205. Kulik E, Cahalan L, Verhoeven M, Cahalan P (1997) Surface modification technology. In: Sudarshan T, Khor K (eds) 10th Proc Int 451. Institute of Materials London C.A. 1997 127:331,898
206. Yu J, Song Z (1996) Hecheng Shuzhi Ji Suliano 13:11 C.A. 1997 127:66,491
207. Ishizaki Y, Yamashita T (1988) JP Patent 63,160,835 Mitsubishi Petrochemical Co C.A. 1988 109:151,111
208. Fujii M, Kitagawa S, Goto S (1986) JP Patent 61,197,640 Mitsubishi Petrochemical Co C.A. 1987 106:51,362
209. Hashimoto Y, Nagaoka Y JP (1993) Patent 05,104,694 Mitsubishi Petrochemical Co C.A. 1993 119:227,470
210. Ko SG, Yoo YS, Yoon YS, Jung BC, Yung HJ, Choi SC, Choi WG (2000) Korea Institute of Science and Technology KR 2,000,039,496 C.A. 2002 136:35,643
211. Osan F, Kulpe F, Kreuder W (1996) EP 694568 Hoechst AG C.A. 1996 124:203,906

Editor: Oskar Nuyken
Received: March 2003

Adv Polym Sci (2004) 167:81–106
DOI: 10.1007/b12305

Functional Macromolecules with Electron-Donating Dithiafulvene Unit

Takashi Uemura · Kensuke Naka · Yoshiki Chujo

Department of Polymer Chemistry, Graduate School of Engineering, Kyoto University, 606-8501, Yoshida, Sakyo-ku, Kyoto, Japan
E-mail: chujo@chujo.synchem.kyoto-u.ac.jp

Abstract This review highlights the recent progress of macromolecules consisted of electron-donating dithiafulvene units. π-Conjugated polymers with the dithiafulvene derivatives, including main-chain and side-chain systems, have been studied with the aim of improving processability of the charge-transfer (CT) complexes and increasing dimensionality of conduction state in the solid state. Such dithiafulvene structures have been applied not only for components of molecular conductor but also π-rich redox active building blocks. Dendritic macromolecules comprised with tetrathiafulvalene (TTF) moiety have accomplished controlled and reversible multi-redox systems. Investigations of interlocked dithiafulvene superstructures toward intelligent molecular devices have been attractive subjects for many scientists. Instead of such dithiafulvene moieties, thioketene dimers have been emerged as promising donor molecules recently, and also utilized for an electroactive polymer and an interesting intramolecular CT system.

Keywords Dithiafulvene · π-Conjugated polymer · Dendrimer · Polyrotaxane · Thioketene dimer

1 Introduction

A dithiafulvene (**1**) can form a 1,3-dithiolium cation by an easy one-electron oxidation, as a result of its electron-donating property. The 1,3-dithiolium ion is an unsaturated five-membered-ring cation in which each sulfur atom contributes a pair of 3π electrons and, consequently, would be expected to show aromatic stability [1, 2]. The chemistry of the dithiafulvene and its derivatives has been a subject of intense interest for the last several decades. Many dithiafulvene derivatives are now well-established as building blocks in the widespread fields of materials chemistry [3–5]. The prominent features of the dithiafulvene are as follows:

1. Dithiafulvene derivatives behave as π-electron donors and form stable charge-transfer complexes and radical ion salts with a wide variety of organic and inorganic acceptor species.
2. Oxidative dimerizations of the dithiafulvenes by both cyclic voltammetry (CV) and chemical method afford dimeric dications (Scheme 1) [6–11].

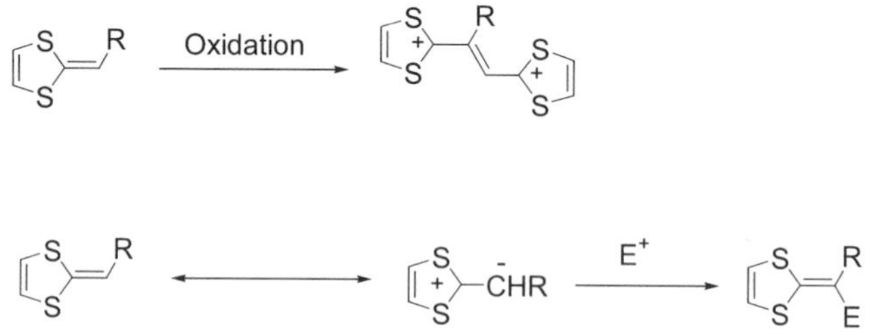

Scheme 1

3. The oxidation potentials can be finely tuned by the attachment of substituents. Delocalized π-electron systems of the dithiafulvenes show low oxidation potentials; however, non-conjugated dithiafulvenes are typically oxidized at between E^{ox}+0.9 and +1.4 V. Electron-sufficient and deficient substituents on the dithiafulvenes also affect the potentials.
4. Dithiafulvenes undergo attack by electrophiles at the exocyclic carbon to give 6-substituted derivatives (Scheme 1).

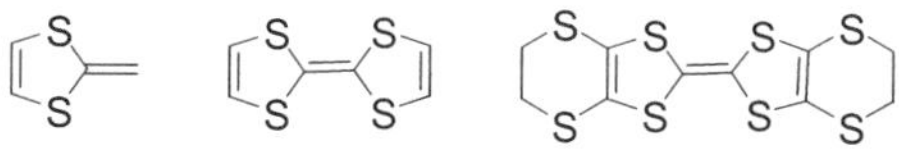

In the chemistry of the dithiafulvenes, especially the focus has been centered on derivatives of tetrathiafulvalene (TTF) (**2**) [13, 14] containing two 1,3-dithiole rings in conjugation, since the discoveries of a high electrical conductivity in a chloride salt of TTF [15] and metallic behavior in the charge-transfer (CT) complex with 7,7,8,8-tetracyano-*p*-quinodimethane (TCNQ) [16, 17]. These crystalline materials are composed of stable segregated stacks of the π-donors and counter-anions. Charge-delocalizations throughout the column structures induce short interplanar distances, giving rise to efficient interactions between neighboring π-molecular orbitals, consequently, a high anisotropic conductivity along the direction of stacking through intermolecular intra-stack migration of aromaticity [18].

Despite the inherent electronic advantages of the dithiafulvenes, the donors have two serious problems. The first one is that their CT complexes and radical cation salts tend to be brittle and unprocessable. This can be improved by the incorporation of the dithiafulvenes into polymeric backbones that are well known for their good processability and film-forming property. Several attempts to incorporate the donor units into main-chain or side-chain of polymers have been achieved to prepare formations of good processable CT complexes [19–25]. The second one is that some dithiafulvene-based conductors, especially TTF, behave as quasi-one-dimensional metals because of their highly-ordered stacks. Such low-dimensional conductors have been predicted by Peierls to undergo a lattice distortion at low temperatures resulting in a metal-insulator transition [26, 27]. To overcome this issue, modifications of the dithiafulvenes to produce new conductors of higher dimensionality have been intensively studied [28–30]. Conjugation-extended donor molecules containing an extended π-framework between the dithiafulvene moieties are expected as promising electron donors for the organic metals, since the extended conjugation decreases the intramolecular Coulomb repulsive energy between the donor units and hence increases the stability of the oxidation states to enhance intramolecular and interstack interactions (increased dimensionality) [31–33]. An increase of the sulfur content in the structure is also effective to achieve the high dimension. Bisethylenedithiotetrathiafulvalene (BEDT-TTF, **3**) [34] contains extra chalcogen atoms, which can participate in interstack as well as intrastack interaction. The peripheral ethylene units of **3** are flexible allowing the molecule to adopt a variety of different conformations. Most BEDT-TTF CT salts are found to be quasi-two-dimensional compounds and some partially oxidized salts have shown ambient-pressure superconductivities up to ca. 12 K [35–37].

A hybrid system between the dithiafulvene derivatives and π-conjugated polymers is interesting. Many kinds of the π-conjugated polymers have been synthesized and studied extensively for a past few decades owing to their linear π-electron delocalization systems extending over a large number of recurrent units [38–41]. The dithiafulvene derivatives and the π-conjugated polymers are representative examples of the main two classes of the organic metals. Although these two classes of the materials differ in many respects, the association of some of their structural characters and properties may contribute to mutual fertilization by development of new materials of both fundamental and technological interest. The conjugated polymers are basically low dimensional conductors, similarly to the TTF CT salts. An incorporation of the donor units into the conjugated polymers would increase the dimensionality of the conduction process in the CT state, due to improved electron mobility along the polymer backbone via π-conjugation, as well as along the stacking direction via π-orbital overlap [28–30]. Furthermore, the donor polymer constitutes a highly polarizable species due to the large number of sulfur atoms in the structure. Therefore, inter- and intramolecular Coulombic repulsions between the charged species would be suppressed, resulting in enhanced stacking CT formations. The processability of the resulting polymer CT salts, needless to say, are better than single CT salts.

The dithiafulvene derivatives have been utilized not only for components of the molecular conductor but also π-rich redox active building blocks [5, 42]. The reversible redox and intermolecular CT properties of the donors are attracting attention in the context of molecular/supramolecular systems which are prototype nanoscale devices. Interlocked and intertwined molecules (e.g., catenanes and rotaxanens) based on the dithiafulvene units could act as chemically- and electrochemically-switches at the molecular level [43–47]. Dithiafulvene derivatives have been also used with a great success as materials with marked nonlinear optical properties [48, 49]. Another interesting application implicates crown-annelated TTFs [50–56], in which the crown ether moiety is located at the dithiole rings, as complexing agents for the recognition of metal ions and for the synthesis of electroactive donor-acceptor diads [57–63]. Development in synthetic dithiafulvene chemistry during the past decade has allowed the preparation of a number of such elaborate molecular systems.

In this review, the recent progresses of functional macromolecules comprised with the dithiafulvene moiety, e.g., dithiafulvene/π-conjugated polymer hybrid systems, dendritic dithiafulvenes, polyrotaxane structures, and intramolecular CTs are highlighted.

1.1 Conjugated Dithiafulvenes in Polymer Main-Chain System

Since the first polymer containing the TTF moiety was prepared via a polycoupling process [19], a lot of attempts to incorporate such donor molecules into a polymeric framework have been carried out. Although some of them showed semiconducting properties upon doping, they were generally disordered or poorly characterizable materials. Progress in synthetic chemistry of the dithiafulvene derivatives during the past decade has allowed the preparation of new conjugated polymeric donor systems, in particular, the donor units incorporated in the conjugated main-chain systems.

Fully conjugated ribbon structure polymers with TTF unit (**4**, **5**) were prepared by Müllen and coworkers [64, 65]. Although **5** showed poor solubility owing to its rigid structure, **4** was soluble in common organic solvents such as dichloromethane and THF. Different approach via processable precursor polymers gave **4** and **5** effectively. For polymer **4**, the highest conductivity (0.55 S/cm) was measured with an iodine content (I_3^-) of 60%. For **5**, the maximum conductivity of 1.5×10^{-2} S/cm was detected at 55% iodine content.

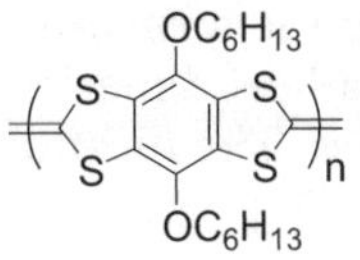

4 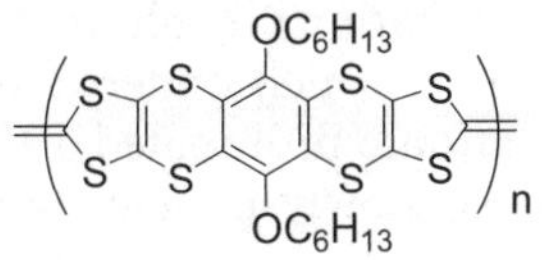 **5**

Yamamoto et al. reported the Ni- and Pd-promoted polycondensations and copolymerizations giving π-conjugated TTF polymers (**6–9**) [66]. Among the polymers, the poly(aryleneethynylene) type polymers **8** and **9** had effectively expanded π-conjugation systems due to C≡C spacer groups and they were active for both chemical and electrochemical oxidation, even though they contained electron-withdrawing C≡C group. Electrical conductivities of **8** and **9** after oxidation with iodine were 4.7×10^{-3} and 2.7×10^{-4} S/cm, respectively. Suzuki coupling reaction between dibrominated TTF derivatives and diboronic acid derivatives afforded conjugated TTF main chain polymers (**10**) [67]. The polymer **10** showed very high solubility owing to the introduction of the alkoxy chains. The color of the films of **10** changed reversibly from yellow to red purple depending on the applied potential in CV. The conductivities of **10** in non-doped state were below 10^{-10} S/cm. After doped with iodine, **10a** and **10b** showed electrical conductivities of 3.8×10^{-7} and 4.9×10^{-7} S/cm, respectively. The relatively low conductivities of **10**, compared with those of other TTF polymers, may result from amorphous structures of **10**.

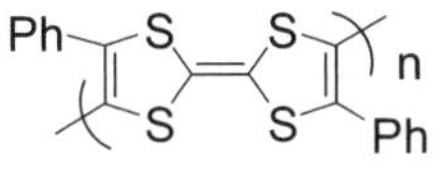

6 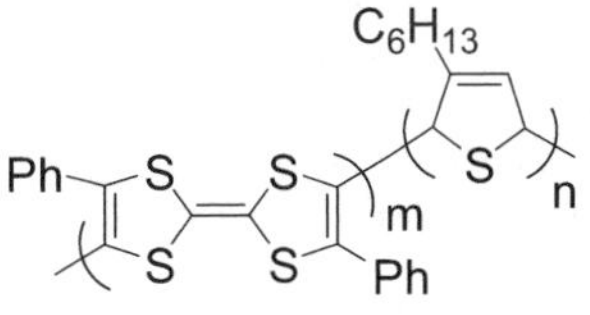 **7** 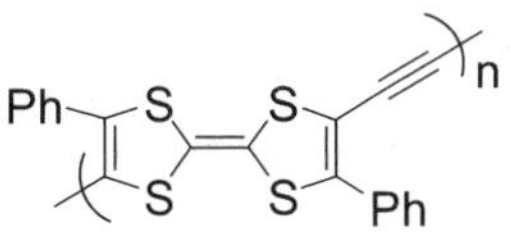 **8**

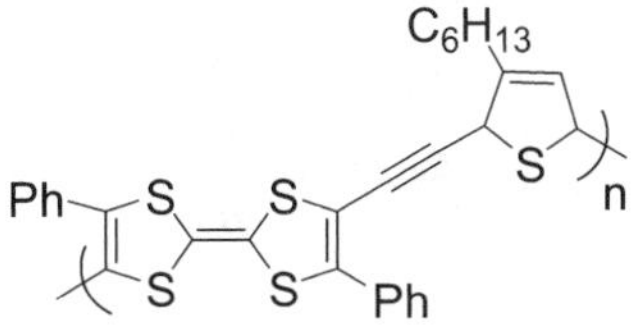

9

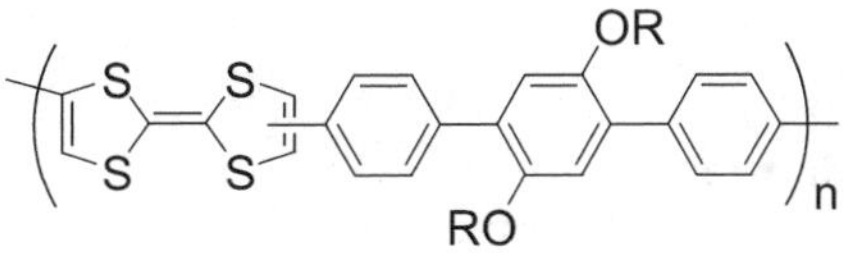

10a; R = C_6H_{13}
10b; R = $C_{12}H_{25}$

The oxidation of di(benzylidene)tetrathiapentalenes in CV resulted in the stepwise formation of linearly extended TTF polymers (**11**) (Scheme 2) [68,

69]. The polymerizations involved as a first step the formation of the monomer radical cations which underwent rapidly radical dimerization reactions to produce dicationic protonated TTF derivatives. The dicationic intermediates deprotonated slowly to stable vinylogous TTF polymers. This new synthetic strategy would offer promising approach to prepare various kinds of TTF polymers. Although the reaction mechanism was investigated in detail, this report did not describe hopeful physical and electrical properties of **11**.

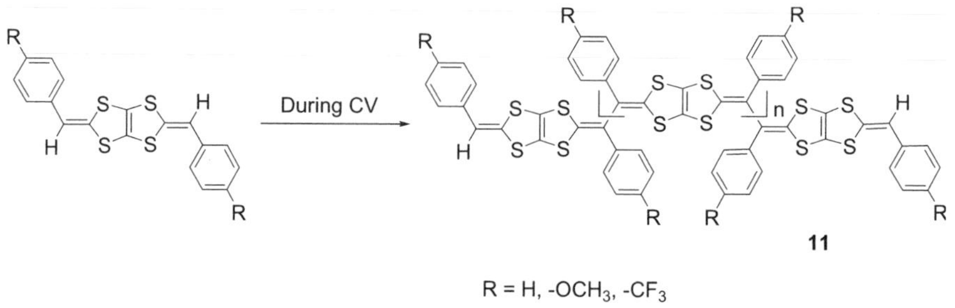

During CV

11

R = H, $-OCH_3$, $-CF_3$

Scheme 2

A series of π-conjugated poly(dithiafulvene)s (**12**) have been prepared by cycloaddition polymerization of aldothioketenes and their alkynethiol tautomers, which were derived from aromatic diynes (Scheme 3) [70–73]. Efficient expansions of π-conjugation systems in the polymers were evident from their UV-vis absorption spectra (Table 1). The polymer **12c** with dodecyl substituent on phenyl moiety exhibited the most extended π-conjugation system [74], owing to increased coplanarity of adjacent units induced by the long alkyl-chain. Absorptions of heteroaromatic polymers (**12d, 12e**) were located at longer wavelengths than those of aromatic polymers (**12a, 12b**),

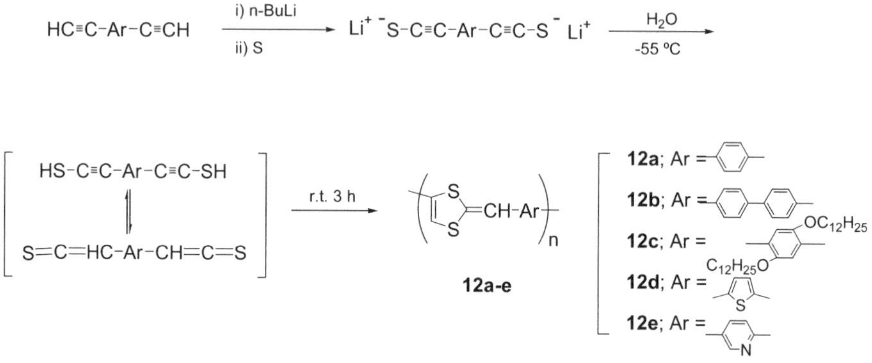

HC≡C–Ar–C≡CH —(i) n-BuLi; ii) S)→ Li^+ $^-$S–C≡C–Ar–C≡C–S$^-$ Li^+ —(H_2O, -55 °C)→

[HS–C≡C–Ar–C≡C–SH ⇌ S=C=HC–Ar–CH=C=S] —(r.t. 3 h)→ **12a-e**

12a; Ar =
12b; Ar =
12c; Ar = ($OC_{12}H_{25}$, $C_{12}H_{25}O$)
12d; Ar =
12e; Ar =

Scheme 3

Table 1 Various properties of poly(dithiafulvene)s **12**

	UV-vis absorption λ_{max} (nm)	Oxidation peak E_{ox} (V)[d]	Conductivity (S/cm)	
			Non-dope	TCNQ complex
12a	398[a]	0.61	3×10^{-7}	2×10^{-4}
12b	379[a]	0.75	-	-
12c	451[b]	0.74	$<10^{-7}$	4×10^{-6}
12d	421[a]	0.78	1×10^{-6}	1×10^{-4}
12e	416 (sh)[a,c]	1.38	2×10^{-4}	3×10^{-4}

[a] Measured in CH_3CN

[b] Measured in $CHCl_3$

[c] Shoulder peak

[d] Measured in CH_3CN solution of 0.1 mol/l $[NEt_4]BF_4$ at 300 mV/s

indicating effective π-conjugation through the heteroaromatic moieties [73]. The CV analyses of **12** showed oxidation peaks for the dithiafulvene unit from 0.6 to 1.4 V. The fairly high oxidation potential of **12e** was caused by an interaction between the electron-sufficient dithiafulvene and the electron-deficient pyridine units [73]. The polymer **12** formed CT complexes with TCNQ (Scheme 4), which were soluble in common organic solvents and

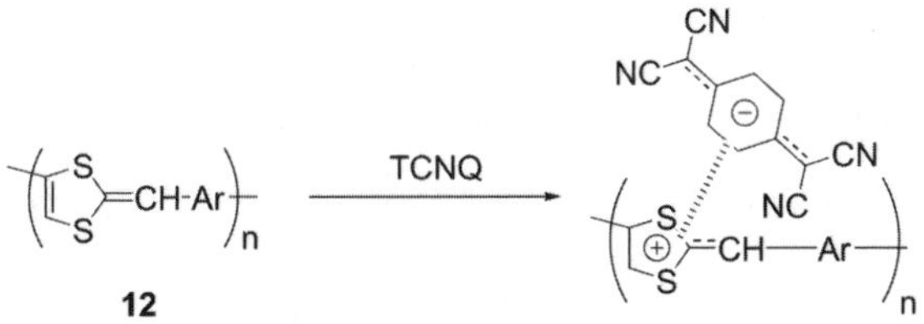

Scheme 4

formed processable films [72]. The stoichiometries of the CT complexes were the dithiafulvene repeating unit/TCNQ=1/1. Electrical conductivities of **12** were improved after formation of the CT complexes (Table 1). For example, the CT complex of **12a** showed conductivity, which was three orders of magnitude greater than that of the uncomplexed polymer. Unexpectedly, the conductivity of **12e** in non-doped state was significantly high [73]. This observation was explained by intramolecular CT effect between the dithiafulvene and the pyridine moieties. An oxidation of **12a** with iodine raised the electrical conductivity of 1×10^{-3} S/cm [71].

Stable colloidal forms of nanocomposites protected with the polymer **12a** have been recently synthesized (Scheme 5) [75]. Gold colloidal particles were formed with narrow size distribution (average size 6 nm) via reduction of $HAuCl_4$ by **12a** due to its electron donating property. The oxidized polymer then protected and stabilized the gold nanoparticles which were stable in

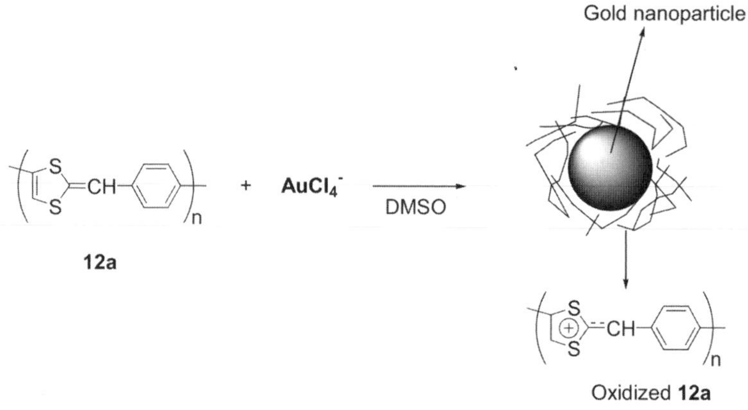

Scheme 5

DMSO without precipitation for more than one month under air. This convenient preparation of the nanoparticle gave new methodology for interesting hybrid systems consisting of inorganic nanoparticles with π-conjugated polymers.

Poly(dithiafulvene)s toward intelligent functional materials have been synthesized [76–78]. An alternating π-conjugated copolymer (**13**) of ferrocene with dithiafulvene showed a unique redox property [76]. Although **13**

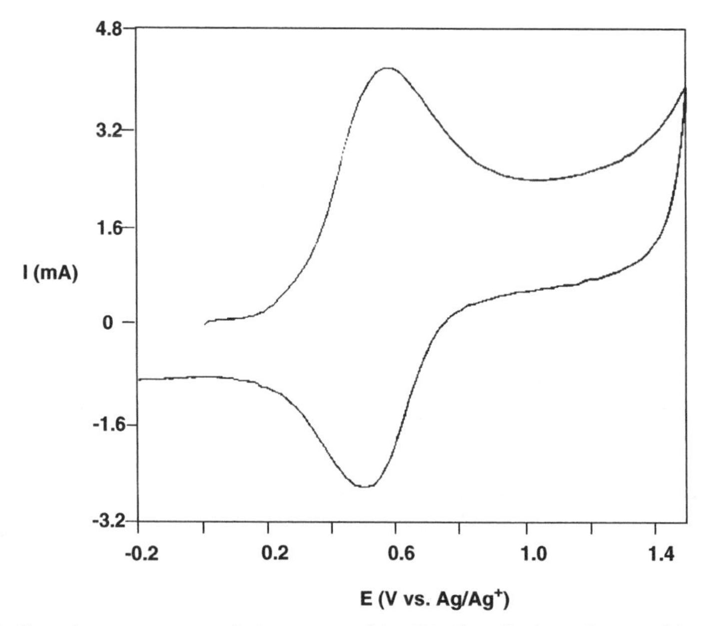

Fig. 1 Cyclic voltammograms of **13**, measured in CH_2Cl_2 solution of 0.2 mol/l tetrabutylammonium hexafluorophosphate at 300 mV/s

had two kinds of redox active sites in the main chain, the cyclic voltammogram showed only single-broad oxidation peak at 0.58 V involving 1.21 electron transfer (Fig. 1), resulting from an effective interaction between two different donors in **13**. Usually, the oxidation of the dithiafulvenes leads to the formation of the dimers during CV scans (Scheme 1). It should be noted that such a dimerization of **13** was not observed during the CV. On repeated cycling, there was no marked change in the trace and no evidence was obtained for the dimer formations due to the high stability of **13** in an oxidative state. This combined redox system of **13** would offer new applications of molecular materials with interesting properties such as unique conducting, magnetic and non-linear-optical (NLO) effects. A poly(dithiafulvene) having 2,2′-bipyridyl unit (**14**) gave a formation of ruthenium complex [77]. Cyclic voltammogram of the complex showed redox peaks, which were characteristic of the dithiafulvene unit and the tris(bipyridyl) ruthenium complex. Alternating donor-acceptor π-conjugated copolymers of the dithiafulvene with cyclodiborazane (**15**) were prepared by hydroboration polymerization [78]. The polymer **15a** showed absorption maximum and fluorescence emissions in chloroform at 415 and 531 nm, respectively. The poly(dithiafulvene) **12a** did not show the emission. The relatively large Stokes shift of **15a** system was explained by effective energy transfer from the dithiafulvene unit to the cyclodiborazane unit in the polymer. All the polymers **13–15** formed soluble CT complexes with TCNQ.

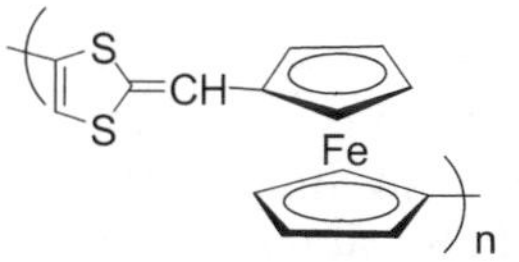

13 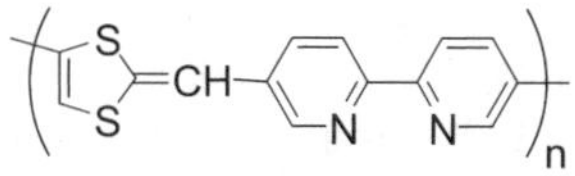 **14**

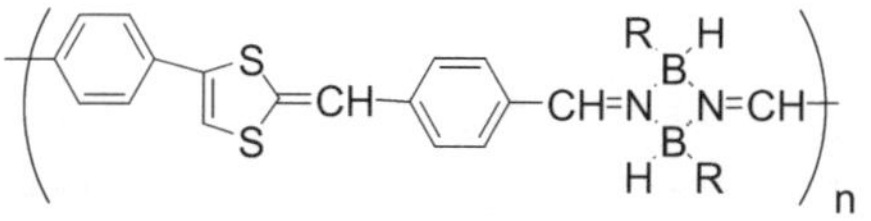

15a; R = Tripyl
15b; R = Mesityl

A new σ-π conjugation system through saturated polymethylene chains has been established very recently [79]. Initial UV-vis absorption study of polyviologens and ab initio calculations suggested a possibility of conjugative effects through saturated polymethylene chains by combination with pyridinium moiety in polymeric frameworks. The polymers **16a–e** with an electron acceptor-donor-acceptor (pyridinium-dithiafulvene-pyridinium) moiety in the π-unit showed the absorptions which were located at significantly longer wavelengths than that of the model compound (**16f**) and shifted gradually to lower energy side as the methylene spacer decreased (Fig. 2). These facts showed that an incorporation of the acceptor-donor-acceptor structure, particularly **16a,** enhanced the conjugative effect through poly-

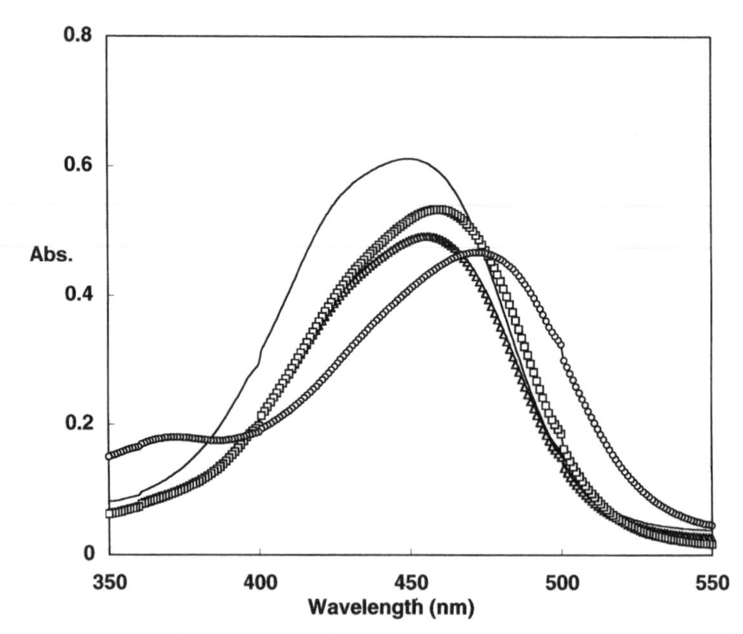

Fig. 2 UV-vis spectra of **16a** (*open circles*), **16c** (*open squares*), **16e** (*open triangles*), and **16f** (*solid line*) in DMSO

methylenes. The electron-donating dithiafulvene units increased electron densities of the adjacent pyridinium moieties by the CT interaction, which would effectively enhance the electronic flows along the polymer chains. Usually polymethylene spacer units inevitably act as interrupters of conjugative effects between π-units, however, use of intramolecular CT effect induced by the dithiafulvene derivative showed efficient σ-π conjugations between the polymethylene chains and the π-units.

16a-e (**a**,**b**,**c**,**d**,**e** for n = 2,3,4,5, and 12, respectively) **16f**

1.2 Incorporation of Donors Into Side-Chains of π-Conjugated Polymers

Similarly to monomeric donors, side-chain donor polymers can afford formation of partially stacking donor moieties, which is necessary for high conductivity, if they achieve a smectic phase. A thiophene derivative con-

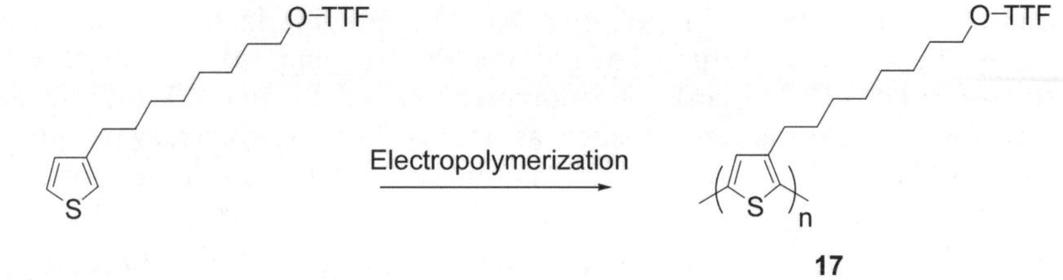
O–TTF
Electropolymerization
S
O–TTF
S
n
17

Scheme 6

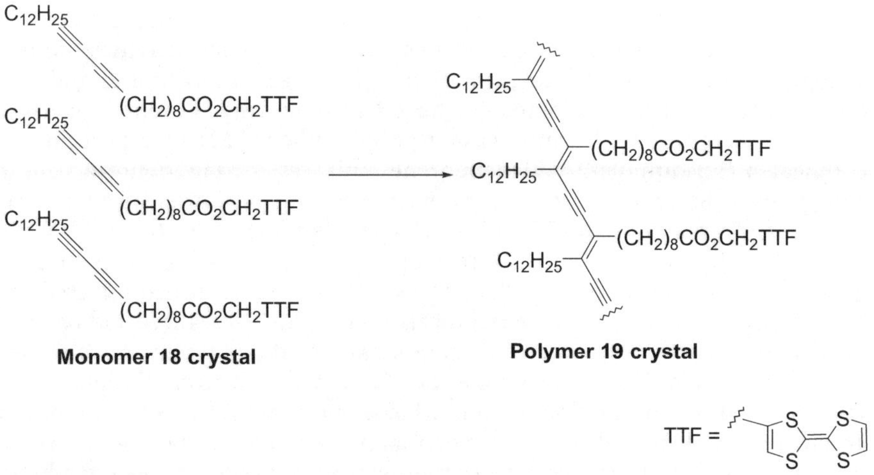
C12H25
(CH2)8CO2CH2TTF
C12H25
(CH2)8CO2CH2TTF
C12H25
(CH2)8CO2CH2TTF
Monomer 18 crystal
C12H25
(CH2)8CO2CH2TTF
C12H25
(CH2)8CO2CH2TTF
C12H25
Polymer 19 crystal
TTF =
S
S
S
S

Scheme 7

taining the TTF moiety covalently attached to the thiophene ring (**17**) was synthesized [80]. Electropolymerization of the thiophene unit of **17** by CV in acetonitrile was unsuccessful because of the scavenging of the thiophene cation radicals by the TTF. In contrast, the polymerization could be achieved in nitrobenzene solutions (Scheme 6), suggesting that the formation of a donor-acceptor complex between the TTF and nitrobenzene decrease its reactivity toward the thiophene cation radicals [80]. A complicated oxidation process of the resultant polymer (**17**) was investigated. A diacetylene containing a TTF substituent (**18**) was synthesized by Shimada et al., which polymerized in the solid state (Scheme 7) [81, 82]. The color of the bulk crystal of **18** changed to blue from light yellow after exposure to UV with a low-pressure mercury lamp or ^{60}Co γ-ray irradiation, indicating solid-state polymerization proceeded via 1,4-addition to yield **19**. A detailed study of the polymerization by ^{60}Co γ-ray irradiation was achieved by gravimetry. The polymer content of the crystals increased with the γ-ray dose up to 250KGy, to about 72% conversion.

Fully conjugated side-chain dithiafulvene polymers have been also prepared [83, 84]. Polymerization of 2-ethynyl-TTF by use of a rhodium catalyst produced a TTF-side chain polymer **20**. The polymerization proceeded smoothly at room temperature. The polymer **20** itself showed a low electrical conductivity ($<1\times10^{-9}$ S/cm). The polymer reacted to give CT complexes with various electron acceptors such as TCNQ, tetrafluorotetracyanoquinodimethane ($TCNQF_4$), dichlorodicyanoquinone (DDQ), and iodine. Among them, the CT complex with TCNQ (TTF unit: TCNQ=5:1) showed the highest conductivity of 2.1×10^{-3} S/cm. The polymers **21** were synthesized by electrolytic oxidation of the corresponding dithiafulvenyl bithiophene monomers [85]. Oxidation potentials of **21** depended on the substituents on the dithiole ring, varying between 0.63 and 0.81 V vs SCE. Electrical conductivities of **21a**, **c**, and **d** were 4.1×10^{-1}, 4.4×10^{-1}, and 1.4 S/cm in the doping state, respectively. The doping states of **21** were stable and the conductivities did not change after storage under air for a month. The high conductivities and stability of **21** might be attributed to the 1,3-dithiole ring. Recently, Skabara and coworkers reported synthesis of a polythiophene (**22**) incorporating π-conjugated 1,3-dithiole-2-ylidenefluorene units as strong donor-acceptor components [86]. The polymer was obtained by chemical or electrochemical oxidation. Intramolecular CT within **22** appeared at 567 nm in the UV-vis spectrum. Photoinduced IR spectroscopy of **22** provided an evidence of long-living photoexcited CT in the polymer. Furthermore, broad infrared active vibration (IRAV) bands were indicative of highly delocalized radical cations, suggesting that the electron-donating site of the polymer was situated within the polythiophene backbone. Conversely, the electron-accepting site was localized within the fluorene unit. Electropolymerization of fluorenes is an efficient tool to synthesize an electroactive conducting polymer. Oxidation of **23** in CV showed a coating of the electrode by yellow-brown insoluble deposit as a polymeric product [87]. IR spectroscopy of the insoluble product exhibited the remains of the dithiole unit in the polymeric matrix. Even though **23** formed a 2:1 CT complex with TCNQ, obtained as shiny black needles, the CT formation of poly **23** was not investigated in this report.

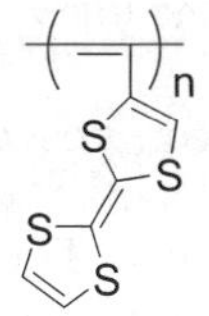

20

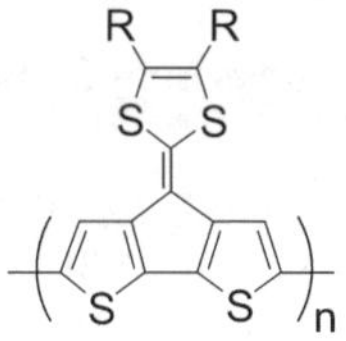

21a; R = H
21b; R = Me
21c; R = benzo
21d; R = SMe
21e; R = CO_2Me
21f; R = $S(CH_2)_9Me$

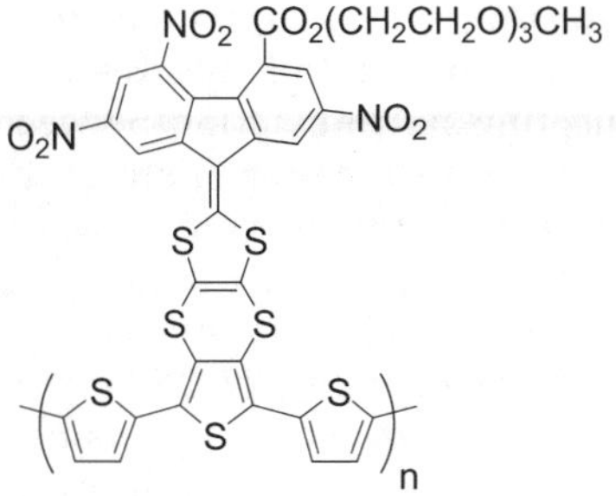

22

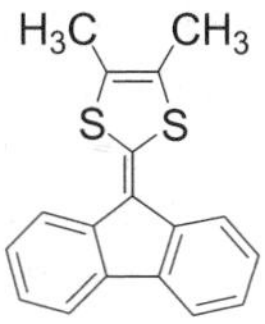

23

A conjugated polymer (**24**) was prepared from 2-benzylidene-4,5-dicyano-1,3-dithiole (Scheme 8) [88]. The UV-vis absorption of **24** gave broad

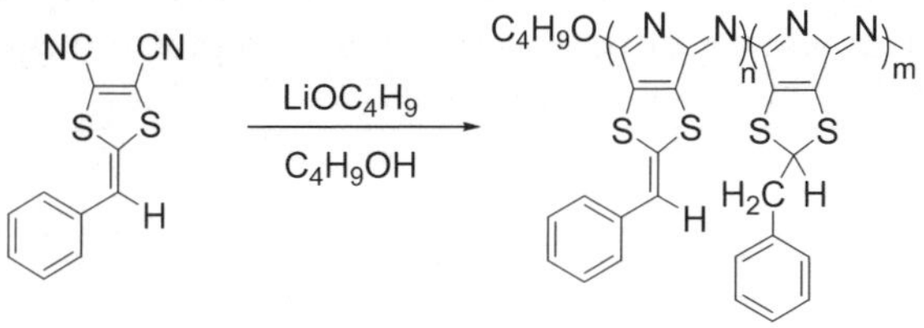

LiOC$_4$H$_9$ / C$_4$H$_9$OH

24

Scheme 8

maxima at 572–580 nm, with tailing past 1000 nm. It is likely that the long absorption tail was due to intramolecular CT structure of **24** and not due to scattering from insoluble polymers. Cast films of **24** showed conductivities in the range 5.5×10^{-6} to 7.5×10^{-7} S/cm. When the film was exposed to iodine vapor, the conductivity increased to 1.1×10^{-4} S/cm.

1.3
Dendritic Macromolecules Comprising Dithiafulvene Units

Dendrimers are sphere macromolecules composed of well-defined branch structures. They also provide regular molecular weights, molecular weight distributions, and chemical structures [89]. The chemistry of the dendrimers has been fascinating subjects in the wide fields of science during this decade. Dendritic macromolecules based on TTF have been synthesized and studied extensively by Bryce and coworkers with an aim of accomplishing controlled and reversible multi-redox systems [90]. The first TTF dendrimer (**25**) was prepared by a convergent strategy, which was reported in 1994 [91]. The dendrimer possessed TTF units at only exterior (surface) of the molecule. The redox behavior of **25** exhibited a characteristic redox behavior typical of the TTF system, involving a simultaneous multi-electron transfer. Formation of the species bearing multiple positive charge produced no significant perturbation, indicating no interaction between the charged TTF units in **25**. The dendrimer **25** was stable when stored below 0 °C, although notable decomposition was reported at room temperature even in the dark and under argon atmosphere after seven days. A dendrimer (**26**) having both TTF and electron-accepting anthraquinone (AQ) units has been synthesized [92]. The solution electrochemistry of **26** was investigated by CV in CH_3CN. Scanning anodically, **26** showed two reversible four-electron oxidation waves to form the radical cation and dication of each of the TTF moieties. Scanning cathodically, two reversible two-electron reduction waves were observed, corresponding to the reduction of each AQ unit producing the radical anion and dianion. This indicated that +16, +8, 0, −4, −8 charged states of **26** was reversibly achieved during the CV scan. The dendrimer **26** showed a very weak broad absorption band in λ=460~750 nm region in CH_3CN, suggesting an intradendrimer CT from TTFs to AQs. Based on the different redox potentials of these donor and acceptor moieties, the degree of CT could be determined to be small (<0.1).

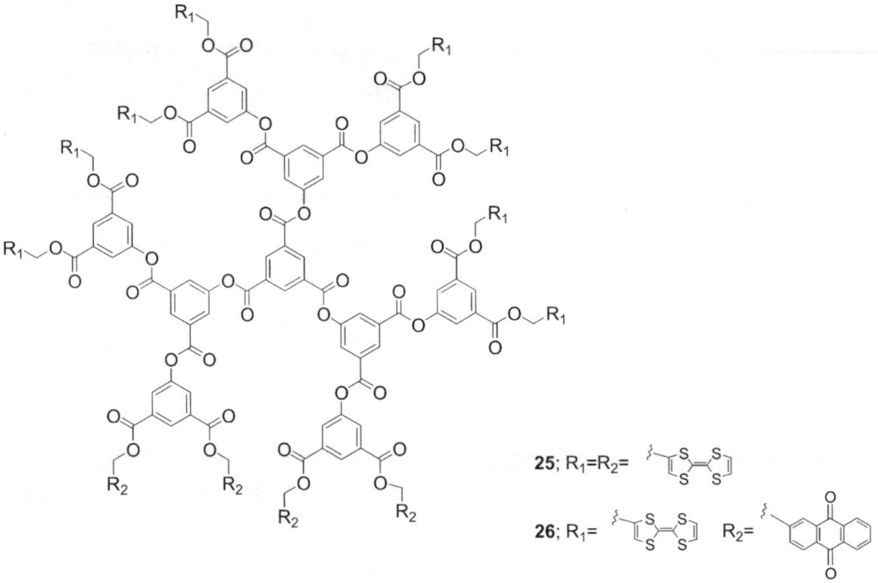

25; R_1=R_2=

26; R_1= R_2=

A TTF-glycol dendrimer (**27**) was reported in 1998 [93, 94]. The dendrimer **27** possessed 21 TTF units and showed high stability against air in contrast to **25**. Thin layer CV techniques of **27** in CH_2Cl_2 showed that all TTF units underwent two single-electron oxidations to generate the +42 oxidation state of the dendrimer. The spectroelectrochemistry of **27** showed partially-oxidized TTF units underwent dimerizations on increasing potential. The dimerized cation radical still remained even in very low concentration. The authors assigned this behavior to an intradendrimer interaction, achieved by the flexible glycol spacers.

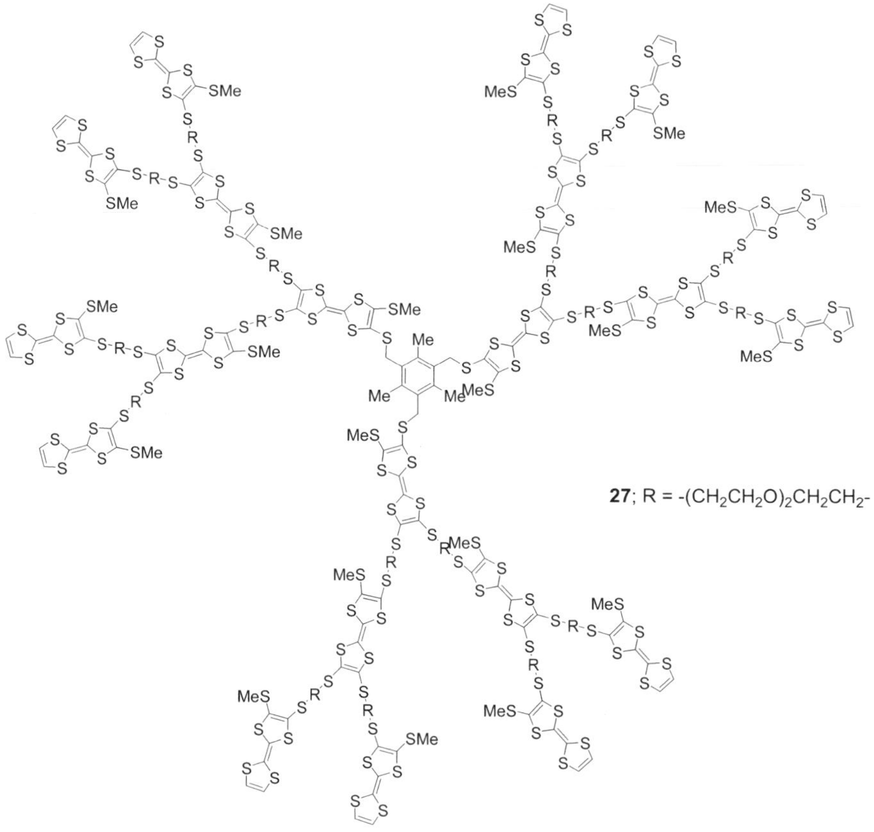

27; R = $-(CH_2CH_2O)_2CH_2CH_2-$

1.4 Polyrotaxane Systems

Mechanically interlocked molecular compounds, including catenanes, rotaxanes, and carceplex, are constituted of molecules composed of two or more components that cannot be separated from each other [95–98]. The development of strategy for achieving controlled self-assembling systems by non-covalent interaction enables one to prepare such attractive compounds for applications in nanoscale molecular devices. The dithiafulvene derivatives are effective electron donors, which are good candidates to form those supramolecular systems with appropriate acceptors by virtue of intermolecular CT interactions. In this chapter, dithiafulvene polymers forming rotaxane structures are especially described.

Stoddart and coworkers reported that a macrocyclic bisviologen acceptor (**28**) acts as a versatile host for π-electron rich systems, such as diphenol methyl ethers, dinaphthol methyl ethers, and aromatic amino acids [99–102]. They also discovered that TTF (**2**) and **28** form a 1:1 complex in which

TTF is located within the cavity of **28** (Scheme 9) [103, 104]. Mixing **2** and **28** in an equimolar proportion in CH_3CN produced an emerald green solution as a result of the appearance of a CT absorption band centered on 854 nm. The association constant K_a of the complex was determined to be

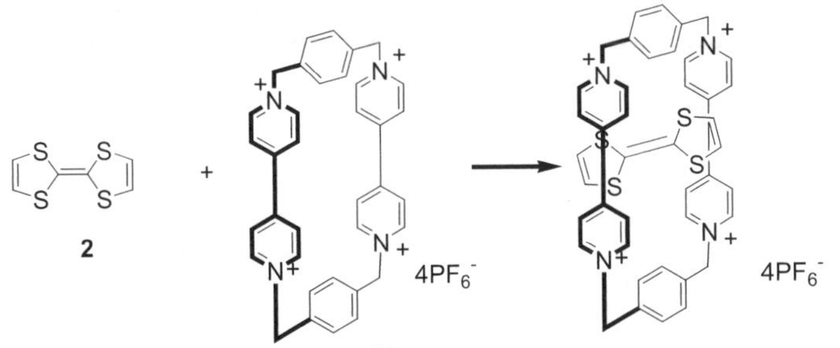

Scheme 9

1000 M^{-1} in CH_3CN and 2600 M^{-1} in acetone. Bryce et al. studied electrochemistry of the inclusion complex in CH_3CN by use of CV [105]. The first oxidation of TTF itself occurred at $E_1^{1/2}$+0.370 V and this value was shifted anodically by 30 mV upon addition of 0.6 equivalents of **28** and by 45 mV upon addition of 1.2 equivalents of **28**. An addition of greater excess of **28** resulted in no further change in the CV. The potentials for second oxidation of TTF remained constant, demonstrating that TTF^{2+} species was not generated in the cavity.

A polyrotaxane **29** possesses two electron-donating sites (TTF and hydroquinone moieties) as stations in the polymer backbone, hence, the incorporated cyclic acceptor **28** moves by external stimuli and possibly two translational isomers (**29a** and **b**) would exist (Scheme 10) [106, 107]. The ratio between two isomers was reported to be very solvent dependent (Table 2), with a preference however for the hydroquinone moiety. In the CV measurement, it was also observed that the cyclic acceptor **28** moved from TTF to hydroquinone moiety along the chain of **29** upon oxidation of the TTF unit.

Table 2 Solvent dependent ratios between **29a** and **b**

Solvent	Isomer **29a** %	Isomer **29b** %
DMSO-d_6	33	67
DMF-d_7	29	71
CD_3CN	12	88
CD_3NO_2	7	93
Acetone-d_6	0	100

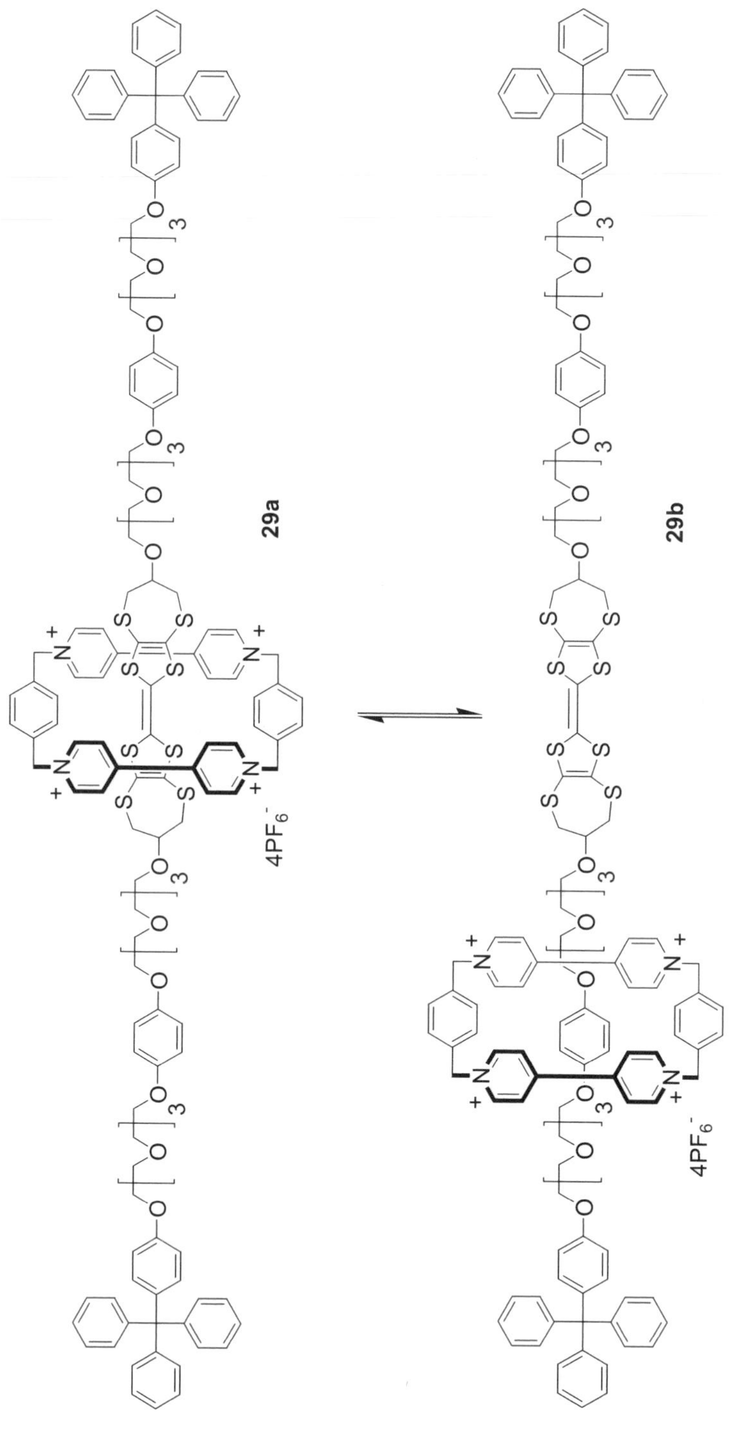

29a

$4PF_6^-$

29b

$4PF_6^-$

Scheme 10

Similarly to TTF, a dithiafulvene **30** was inserted into the cavity of **28** creating a 1:1 complex as a pseudorotaxane formation by a CT interaction between **28** and **30** (Scheme 11 upper) [108]. The complex in DMSO showed a

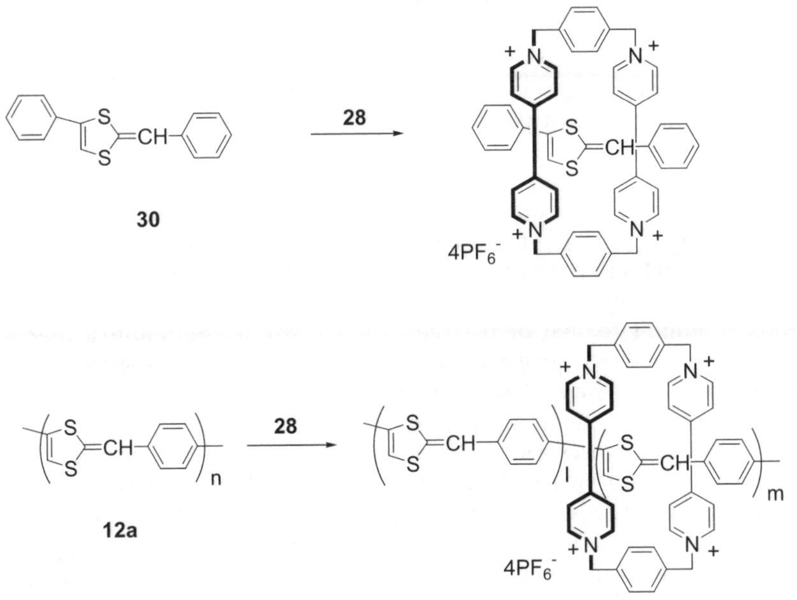

Scheme 11

charge-transfer absorption band with a peak at 616 nm in the UV-vis measurement. Spectrophotometric dilution analysis performed on DMSO solution at this wavelength at 25 °C yielded an association constant K_a=51 M^{-1}. This value was lower than that obtained for the TTF-**28** complex due to the weaker electron donating ability of **30** (E_{pa}=0.40 V) than that of TTF (E_{pa}=0.39 and 0.81 V). The π-conjugated polymer **12a** (n=10.3, M_n=2320) gave a pseudopolyrotaxane formation with **28** (Scheme 11 bottom) in a similar manner to **30** [108]. DMSO solution of **12a** and **28** showed an absorption which was diagnostic of CT complexation between the dithiafulvene moiety in **12a** and **28**. The ^{1}H NMR study suggested that the cyclic acceptor **28** was located on 3.2 units of the repeating units in the polymer. After complexation with **28**, the oxidation potential of **12a** shifted anodically to 0.83 V. The uncomplexed **12a** exhibited the electrical conductivities of 3.4×10^{-7} S/cm. The polymer CT complex with **27** in the ratio of the dithiafulvene unit of **12a** to **28**=10:2 had a conductivity of 7.6×10^{-4} S/cm. This improved conductivity arose from the effective CT interaction between the dithiafulvene unit and the cyclic acceptor **28**. There have been very limited examples that reported on π-conjugated polymers with rotaxane moiety. Such hybrid system should show interesting electronic properties and will find the application in the macromolecular, supramolecular, and material fields of chemistry.

1.5
Conjugated Polymers with Thioketene Dimer Unit

There is a long standing interest in the chemistry and the properties of cyclic compounds containing sulfur atom in modern material chemistry due to their redox chemistry. In particular, the focus has been on dithiole derivatives, e.g., dithiafulvenes and tetrathiafulvalenes, since the finding of metallic conductivity and low temperature superconductivity in radical cation salts. The quite low oxidation potentials of 1,4-dithiin compounds have been reported, recently [109]. On the other hand, thioketene dimers (2,4-bis(alkylidene)-1,3-dithietane) have been known for more than 100 years and synthesized by various methods [110–115]. The structure of these dimer compounds is similar to that of the redox-active sulfur compounds; therefore, the potential electronic property of the thioketene dimer moiety is considerably attractive with the aim of application to a new and better π-donor.

We have reported the first electroactivity of a thioketene dimer compound [116]. The CV measurement of 2,4-dibenzylidene-1,3-dithietane (**31**), which was prepared by a basic dimerization of phenylthioketene derived from benzyltriphenylphosphonium chloride, showed irreversible two-step oxidation peaks at 0.25 and 0.61 V vs Ag/Ag^+, indicating that **31** acts as a stronger electron donor than 2,6-bisphenyl-1,4-dithiafulvene (**30**) and TTF (**2**). The dimer (**31**) can form a 1:1 CT complex with TCNQ in DMSO. Cycloaddition polymerization of bisthioketene derived from *p*-xylenebis(triphenylphosphonium chloride) gave a π-conjugated polymer (**32**) with thioketene dimer unit in the main chain (Scheme 12). This polymer was the first polymer contain-

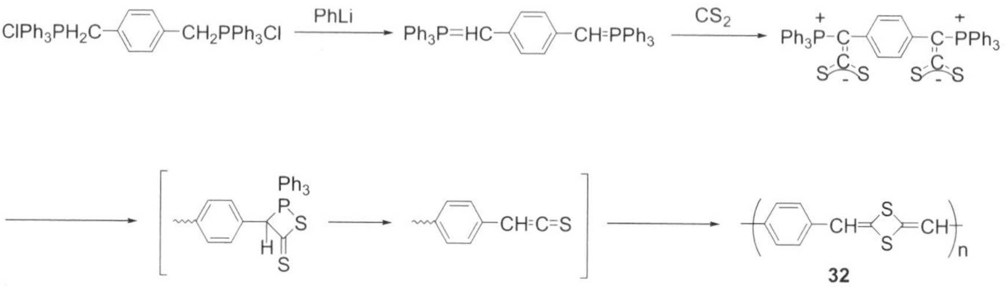

ClPh$_3$PH$_2$C–C$_6$H$_4$–CH$_2$PPh$_3$Cl —PhLi→ Ph$_3$P=HC–C$_6$H$_4$–CH=PPh$_3$ —CS$_2$→ Ph$_3$P$^+$–C(CS$_2^-$)–C$_6$H$_4$–C(CS$_2^-$)–P$^+$Ph$_3$ → [~C$_6$H$_4$–C(H)(PPh$_3$)(S)C=S → ~C$_6$H$_4$–CH=C=S] → 32 (n)

Scheme 12

ing thioketene dimer unit in the structure. Although the molecular weight of **32** was relatively low, UV-vis absorption of **32** showed a red-shift compared with that of **31**, indicating an effective expansion of π-conjugation system in **32**. The cast film of **32** gave irreversible two-step oxidation peaks at 0.39 and around 0.90 V vs Ag/Ag^+ in CV measurement. In DMSO, **32** reacted with TCNQ to produce a CT complex, which had an electrical conductivity of

9.5×10^{-5} S/cm at room temperature. The polymer **32** doped with iodine showed the conductivity of 1.8×10^{-3} S/cm.

31

A series of silyl and disilyl substituted thioketene dimers (2,4-diylidene-1,3-dithietane), including polymers, have been synthesized (**33**, **34**) [117]. HOMO-LUMO calculation by PM3 semi-empirical molecular orbital method predicted an intramolecular CT from the thioketene dimer to the Si-Si bond in disilyl thioketene dimer (**33a**) and no electronic interaction between the

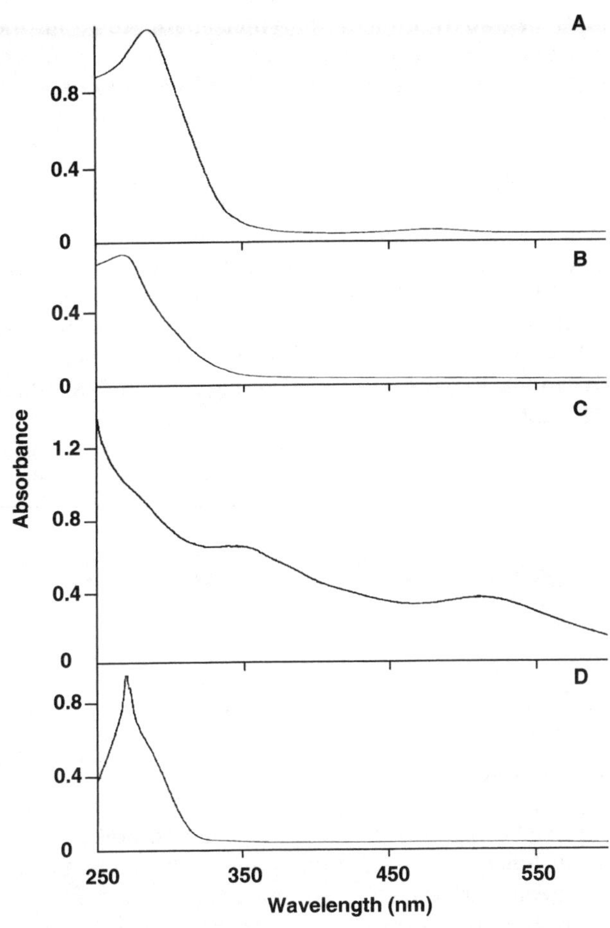

Fig. 3A–D UV-vis spectra of thioketene dimers: **A 33a** in $CHCl_3$; **B 33b** in $CHCl_3$; **C 34a** in CH_3CN; **D 34b** in CH_3CN

Si atom and the thioketene dimer in **33b**. UV-vis absorption measurements of **33** supported the calculation (Fig. 3). The spectrum of **33a** showed small absorption at 463 nm, however, **33b** shows no absorption in this region in $CHCl_3$ (Fig. 2). In addition, the absorption in the visible region of **33a** moved to shorter wavelength, with increasing the solvent polarity (λ_{max}=446 nm in CH_3CN and 455 nm in acetone). This solvent-dependent shift is characteristic of negative-solvatochromism, so that the peak for **33a** is clearly caused by the intramolecular CT interaction between the Si-Si and the thioketene dimer unit. A dimer compound **33b** showed no interaction. Cycloaddition polymerizations of bisthioketenes gave new polymers **34**. The polymer **34a** showed a very large CT absorption (515 nm in CH_3CN) at longer wavelength compared with **33a** (Fig. 2). A negative-solvatochromic behavior of the CT band of **34a** was observed (λ_{max}=488 nm in H_2O, 495 nm in MeOH, and 519 nm in acetone). Sunlight exposure decreased the intensity of the CT absorption of **34a**, suggesting that transformation of the Si-Si unit to a siloxane (Si-O-Si) and/or degradation of the polymer chain led to an inefficient CT interaction. In contrast to **34a**, the polymer **34b** showed no absorption in the visible range (Fig. 2). These facts indicate that the combination between the Si-Si and the thioketene dimer units is necessary to achieve the effective intramolecular CT interaction. It is known that low ionization potential (I_p) of a Si-Si bond leads to electronic delocalization within the σ-bonded framework and effective electron donating property [118–120]. Electron delocalization between the Si-Si σ bonds and π systems (σ-π conjugation) has been also conclusively established [121–124]. Interestingly, the study of compounds **33** and **34** showed that direct attachment of a strong donor such as the thioketene dimer to the Si-Si bond afforded intramolecular CT from the dimer to the Si-Si moiety, even though the Si-Si unit has been regarded as an electron donor so far. Expansion of σ-π conjugation in **34a** enhanced the CT interaction efficiently.

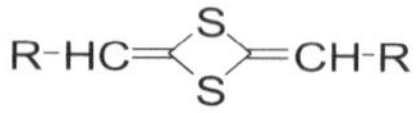

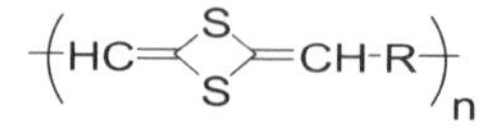

R-HC=C(S)2C=CH-R

33a; R = $-Si_2Me_5$
33b; R = $-SiMe_3$

-(HC=C(S)2C=CH-R)-$_n$

34a; R = $-SiMe_2-SiMe_2-$

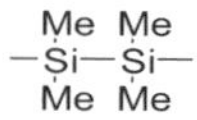

34b; R = $-SiMe_2-$

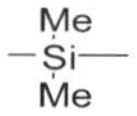

2
Summary and Outlook

The recent approaches toward the functional macromolecules in the dithiafulvene-based systems have been highlighted. The conjugated poly-donor molecules have been synthesized to combine the processability with the significant electron-donating properties of the dithiafulvene systems. Some of them formed CT complexes with organic acceptors, such as TCNQ, achieving higher conductivities and conducting dimensionalities. The dendritic macromolecules including TTF moieties accomplished controlled and reversible multi-redox systems. The high electron-donating abilities of the dithiafulvenes have been exploited in a number of the supramolecular structures, especially catenane and rotaxane systems, to be utilized for sensor, molecular switches and devices. This review also describes that the electronic properties of the thioketene dimers was hopeful, similarly to the dithiafulvene derivatives. The electron-donating ability of the thioketene dimer was revealed to be fairly strong and was utilized for the various unique electronic systems.

References

1. Breslow DS, Skolnik H (1966) Multi-sulfur and sulfur and oxygen five and six membered heterocycles. Part I. Interscience, New York
2. Prinzbach H, Futterer E (1966) Adv Heterocycl Chem 7:39
3. Hansen TK, Becher J (1993) Adv Mater 5:289
4. Bryce MR (1999) Adv Mater 11:11
5. Bryce MR (2000) J Mater Chem 10:589
6. Kirmse W, Horner L (1958) Liebigs Ann Chem 614:4
7. Mayer K, Kröber H (1974) J Prakt Chem 316:967
8. Cava MP, Lakshmikantham MV (1980) Heterocycl Chem 17:S39
9. Schöberl U, Salbeck J, Daub J (1992) J Adv Mater 4:41
10. Lorcy D, Charlier R, Robert A, Tallec A, Le Maguerés P, Ouahab L (1995) J Org Chem 60:2443
11. Hapiot P, Lorcy D, Tallec A, Carlier R, Robert A (1996) J Phys Chem 100:14,823
12. Lakshmikantham MV, Cava MP (1981) J Org Chem 46:3246
13. Wudl F, Smith G, Hufnagel EJ (1970) J Chem Soc, Chem Commun 1453
14. Hünig S, Kisslich G, Sheutzow D, Zahradnik R, Carsky P (1971) Int J Sulfur Chem, Part C 109
15. Wudl F, Wobschall D, Hunfnagel EJ (1972) J Am Chem Soc 94:671
16. Ferraris J, Cowan DO, Walatka V, Perlstein JH (1973) J Am Chem Soc 95:948
17. Coleman LB, Cohen MJ, Sandman DJ, Yamaguchi G, Garito AF, Heeger AJ (1973) Solid State Commun 12:1125
18. Perlstein JH (1977) Angew Chem Int Ed Engl 16:519
19. Ueno Y, Masuyama Y, Okawara M (1975) Chem Lett 603
20. Pittman CU Jr, Narita M, Liang YF (1976) Macromolecules 9:360
21. Hertler WR (1976) J Org Chem 41:1412
22. Pittman CU Jr, Liang YF, Ueda M (1979) Macromolecules 12:541
23. Mulvaney JE, Chang DM (1980) Macromolecules 12:240
24. Koβmehl G, Rohde M (1982) Makromol Chem 183:2084
25. Mulvaney JE, Figueroa FR, Evans SB, Osorio F (1989) J Polym Sci Part A Polym Chem 27:971
26. Peierls RE (1955) Quantum theory of solids. Oxford University Press, London
27. Peierls RE (1964) Quantum chemistry of solids. Clarendon Press, New York
28. Adam M, Müllen K (1994) Adv Mater 6:439
29. Bryce MR (1995) J Mater Chem 5:1481
30. Roncali J (1997) J Mater Chem 7:2307
31. Garito AF, Heeger AJ (1974) Acc Chem Res 7:232
32. Torrance JB (1979) Acc Chem Res 12:79
33. Wudl F (1984) Acc Chem Res 7:227
34. Mizuno M, Garito AF, Cava MP (1978) J Chem Soc Chem Commun 18

35. Kini AM, Geiser U, Wang HH, Carlson KD, Williams JM, Kwok WK, Vandevoot KG, Thompson JE, Stupka DL, Jung D, Whangbo MH (1990) Inorg Chem 29:2555
36. Williams JM, Kini AM, Wang HH, Carlson KD, Geiser U, Montgomery LK, Pyrka GJ, Watkins DM, Kommers JK, Boryschuk SJ, Crouch AV, Kwok WK, Schirber JE, Overmyer DL, Jung D, Whangbo MH (1990) Inorg Chem 29:3274
37. Komatsu T, Nakamura T, Matsukawa N, Yamochi H, Saito G, Ito H, Ishiguro T, Kusunoki M, Sakaguchi K (1992) Solid State Commun 80:101
38. Skotheim TA, Elsenbaumer RL, Reynolds JR (1998) Handbook of conducting polymers, 2nd edn. Marcel Dekker, New York
39. McQuade DT, Pullen AE, Swager TM (2000) Chem Rev 100:2573
40. Prasad PN, William DJ (1991) Introduction to nonlinear optical effects in molecules and polymers. Wiley, New York
41. Zyss J (1994) Molecular nonlinear optics: materials, physics, and devices. Academic Press, Boston
42. Nielsen MB, Lomholt C, Becher J (2000) Chem Soc Rev 29:153
43. Asakawa M, Ashton PR, Balzani V, Credi A, Hamers C, Mattersteig G, Montaliti M, Shipway AN, Spencer N, Stoddart JF, Tolley MS, Venturi M, White AJP, Williams DJ (1998) Angew Chem Int Ed Engl 37:333
44. Ashton PR, Balzani V, Becher J, Credi A, Fyfe MCT, Mattersteig G, Menzer S, Nielsen MB, Raymo FM, Stoddart JF, Venturi M, Williams DJ (1999) J Am Chem Soc 121:3951
45. Nielsen MB, Nielsen SB, Becher J (1998) Chem Commun 475
46. Nielsen MB, Hansen JG, Becher J (1999) Eur J Org Chem 2807
47. Wolf R, Asakawa M, Ashton PR, Gómez-López M, Hamers C, Menzer S, Parsons IW, Spencer N, Stoddart JF, Tolley MS, Williams DJ (1998) Angew Chem Int Ed Engl 37:975
48. Andreu R, De Lucas AI, Garín J, Martín N, Sánchez L, Seoane C (1997) Synth Met 86:1817
49. De Lucas AI, Martín N, Orduna J, Sánchez L, Seoane C, Andreu R, Garín J, Orduna J, Alcalá R, Villacampa B (1998) Tetrahedron 54:4665
50. Hansen TK, Jørgensen T, Stein PC, Becher J (1992) J Org Chem 57:6403
51. Gasiorowski R, Jørgensen T, Møller J, Hansen TK, Pietraszkiewicz M, Becher J (1992) Adv Mater 4:568
52. Dieing R, Morisson V, Moore AJ, Goldenberg LM, Bryce MR, Raoul JM, Petty MC, Garín J, Savirón M, Lednev IK, Hester RE, Moore JN (1996) J Chem Soc Perkin Trans 2 1587
53. Le Derf F, Mazari M, Mercier N, Levillain E, Richomme P, Becher J, Garín J, Orduna J, Gorgues A, Sallé M (1999) Chem Commun 1417
54. Moore AJ, Goldenberg LM, Bryce MR, Petty MC, Monkman AP, Marenco C, Yarwood J, Joyce MJ, Port SN (1998) Adv Mater 10:395
55. Fujihara H, Nakai H, Yoshihara M, Maeshima T (1999) Chem Commun 737
56. Liu H, Echegoyen L (1999) Chem Commun 1493
57. de Miguel P, Bryce MR, Goldenberg LM, Beeby A, Khodorkovsky V, Shapiro L, Niemz A, Cuello AO, Rotello V (1998) J Mater Chem 8:71
58. Scheib S, Cava MP, Baldwin JW, Metzger RM (1998) J Org Chem 63:1198
59. Tsiperman E, Regev T, Becker JY, Bernstein J, Ellern A, Khodorkovsky V, Shames A, Shapiro L (1999) Chem Commun 1125
60. Prato M, Maggini M, Giacometti C, Scorrano G, Sandona G, Farnia G (1996) Tetrahedoron 52:5221
61. Martín N, Sánchez L, Seoane C, Andreu R, Garín J, Orduna J (1996) Tetrahedron Lett 37:5979
62. Llacay J, Veciana J, Vidal-Gancedo J, Bourdelande L, González-Moreno R, Rovira C (1998) J Org Chem 121:3951
63. Simonsen KB, Konovalov VV, Konovalova TA, Kawai T, Cava MP, Kispert LD, Metzger RM, Becher J (1999) J Chem Soc, Perkin Trans 2:657
64. Frenzel S, Arndt S, Gregorious RM, Müllen K (1995) J Mater Chem 5:1529
65. Frenzel S, Baumgarten M, Müllen K (2001) Synth Met 118:97

66. Yamamoto T, Shimizu T (1997) J Mater Chem 7:1967
67. Tamura H, Watanabe T, Imanishi K, Sawada M (1999) Synth Met 107:19
68. Salhi F, Müller H, Divisia-Blohorn B (1999) Synth Met 101:75
69. Hapiot P, Salhi F, Divisia-Blohorn B, Müller H (1999) J Phys Chem A 103:11,221
70. Naka K, Uemura T, Chujo Y (1998) Macromolecules 31:7570
71. Naka K, Uemura T, Chujo Y (1999) Macromolecules 32:4641
72. Naka K, Uemura T, Chujo Y (2000) Polym J 32:435
73. Naka K, Uemura T, Chujo Y (2000) Macromolecules 33:4733
74. Naka K, Uemura T, Chujo Y. Unpublished results
75. Zhou Y, Itoh H, Uemura T, Naka K, Chujo Y (2001) Chem Commun 613
76. Naka K, Uemura T, Chujo Y (2000) Macromolecules 33:6965
77. Naka K, Uemura T, Chujo Y (2001) J Polym Sci, Part A Polym Chem 39:4083
78. Naka K, Umeyama T, Chujo Y (2000) Macromolecules 33:7467
79. Naka K, Uemura T, Chujo Y (2003) Bull Chem Soc Jpn (in press)
80. Thobie-Gautier C, Gorgues A, Jubault M, Roncali J (1993) Macromolecules 26:4094
81. Shimada S, Masaki A, Hayamizu K, Matsuda H, Okada S, Nakanishi H (1997) Chem Commun 1421
82. Shimada S, Masaki A, Matsuda H, Okada S, Nakanishi H (1998) Mol Cryst Liq Cryst Sci Technol, Sect A 315:385
83. Shimizu T, Yamamoto T (1999) Chem Commun 515
84. Shimizu T, Yamamoto T (1999) Inorg Chim Acta 296:278
85. Yamashita Y, Tanaka S, Kozaki M (1992) J Chem Soc Chem Commun 1137
86. Skabara PJ, Serebryakov IM, PerepichkaIf IF, Sariciftci NS, Neugebauer H, Cravino A (2001) Macromolecules 34:2232
87. Rault-Berthelot J, Rozé C (1998) Synth Met 93:103
88. Rixman MA, Sandman DJ (2000) Macromolecules 33:248
89. Newkome GR, Moorefield CN, Vögtle F (1996) Dendritic molecules: concepts, synthesis, perspectives. VCH, Weinheim
90. Bryce MR, Devonport W, Goldenberg LM, Wang C (1998) Chem Commun 945
91. Bryce MR, Devonport W, Moore AJ (1994) Angew Chem Int Ed Engl 33:1761
92. Bryce MR, de Miguel P, Devonport W (1998) Chem Commun 2565
93. Christensen CA, Goldenberg LM, Bryce MR, Becher J (1998) Chem Commun 509
94. Christensen CA, Bryce MR, Becher J (2000) Synthesis 12:1695
95. Fyfe MCT, Stoddart JF (1997) Acc Chem Res 30:393
96. Jäger R, Vögtle F (1997) Angew Chem Int Ed Engl 36:930
97. Sauvage JP, Dietrich-Buchecker CO (1999) Molecular catenanes, rotaxanes and knots. VCH-Wiley, Weinheim
98. Raymo FM, Stoddart JF (1999) Chem Rev 99:1643
99. Odell B, Reddington MV, Slawin AMZ, Spencer N, Stoddart JF, Williams DJ (1988) Angew Chem Int Ed Engl 27:1547
100. Bühner M, Geuder W, Gries WK, Hünig S, Koch M, Poll T (1988) Angew Chem Int Ed Engl 27:1553
101. Goodnow TT, Reddington MV, Stoddart JF, Kaifer AE (1991) J Am Chem Soc 113:4335
102. Bernardo AR, Stoddart JF, Kaifer AE (1992) J Am Chem Soc 114:10,624
103. Phlip D, Slawin AMZ, Spencer N, Stoddart JF, Williams D (1991) J Chem Soc Chem Commun 1584
104. Ashton PR, Balzani V, Becher J, Credi A, Fyfe MCT, Mattersteig G, Menzer S, Nielsen MB, Raymo FM, Stoddart JF, Venturi M, Williams DJ (1999) J Am Chem Soc 121:3951
105. Devonport W, Blower MA, Bryce MR, Goldenberg LM (1997) J Org Chem 62:885
106. Ashton PR, Bissell RA, Spencer N, Stoddart JF, Tolley MS (1992) Synlett 923
107. Anelli PL, Asakawa M, Ashton PR, Bissell RA, Clavier R, Górski R, Kaifer AE, Langford SJ, Mattersteig G, Menzer S, Philp D, Slawin AMZ, Spencer N, Stoddart JF, Tolley MS, Williams DJ (1997) Chem Eur J 3:1113
108. Naka K, Uemura T, Chujo Y (2002) Bull Chem Soc Jpn 75:2053
109. Nishinaga T, Wakamiya A, Komatsu K (1999) Chem Commun 777

110. Raasch MS (1970) J Org Chem 35:3470
111. Raasch MS (1972) Chem Commun 577
112. Pedersen CT, Oddershede J, Sabin JR (1981) J Chem Soc Perkin Trans 2:1062
113. Purrello G, Fiandaca P (1976) J Chem Soc Perkin Trans 1:692
114. Selzer T, Rappoport Z (1996) J Org Chem 61:7326
115. Schönberg A, Frese E, Brosowski K (1962) Chem Ber 95:3077
116. Uemura T, Naka K, Gelover-Santiago A, Chujo Y (2001) Macromolecules 34:346
117. Naka K, Uemura T, Chujo Y (2001) J Am Chem Soc 123:6209
118. Fuchigami T (1998) The chemistry of organosilicon compounds, vol 2. Wiley, New York, chap 20
119. Geniès EM, Omar FE (1983) Electrochim Acta 28:541
120. Yoshida J (1994) Topics in current chemistry, vol 170. Electrochemistry V. Springer, Berlin Heidelberg New York
121. Ishikawa M (1978) Pure Appl Chem 50:11
122. Ishikawa M, Kumada M (1981) Adv Organomet Chem 19:51
123. Sakurai H (1980) J Organomet Chem 200:261
124. Iwahara T, Hayase S, West R (1990) Macromolecules 23:129

Editor: Kwang-Sup Lee
Received: November 2002

Adv Polym Sci (2004) 167:107–135
DOI: 10.1007/b12306

Synthesis of Polyisobutylene-Based Block Copolymers with Precisely Controlled Architecture by Living Cationic Polymerization

Younghwan Kwon[1, 2] · Rudolf Faust[1]

[1] Polymer Science Program, Department of Chemistry, University of Massachusetts Lowell, One, University Avenue, Lowell, MA, 01854, USA
E-mail: Rudolf_Faust@uml.edu

Present address:
[2] Department of Chemical Engineering, Daegu University, Gyeongsan, Gyeongbuk, 712-714, Korea
E-mail: y_kwon@webmail.daegu.ac.kr

Abstract During the last two decades we have witnessed the discovery and development of living cationic polymerization. As a result of persistent growth in this field, today living cationic polymerization is an equal counterpart of anionic living polymerization although the latter has a considerably longer history. Living cationic polymerization is an indispensable tool for the preparation of a wide variety of homo-, block-, graft- and functional polymers. This chapter reviews recent efforts in cationic macromolecular engineering with emphasis on fundamental concepts and advanced technologies for polyisobutylene block copolymer synthesis. For block copolymer synthesis by the simple sequential monomer addition technique rationalization is given for the selection of polymerization conditions and monomer addition order. The application of non-(homo)polymerizable monomers such as 1,1-diarylethylenes and furan analogues, coupled with changes in the synthetic approaches for various block copolymer architectures, is described in terms of three categories; capping, coupling, and living coupling reactions. These appear to be versatile synthetic tools for numerous linear di- and triblock copolymers, and block copolymers with non-linear architectures. Finally, current developments in the combination of different polymerization mechanisms are illustrated.

Keywords Living cationic polymerization · Polyisobutylene · Block copolymers · Macromolecular architecture · Combination of polymerization mechanism

List of Abbreviations and Symbols

AA′B	Poly[(A)-*s*-(A′)-*s*-(B)] three-arm star-block copolymer
A_2B	Poly[$(A)_2$-*s*-(B)] three-arm star-block copolymer
A_2B_2	Poly[$(A)_2$-*s*-$(B)_2$] four-arm star-block copolymer
ABC linear	Poly[(A)-*b*-(B)-*b*-(C)] linear triblock copolymer
ABC star	Poly[(A)-*s*-(B)-*s*-(C)] three-arm star-block copolymer
$AcNMe_2$	*N*,*N*-Dimethylacetamide
AROP	Anionic ring-opening polymerization
BDPEP	2,2-Bis[4-(1-phenylethenyl)phenyl]propane
BDTEP	2,2-Bis[4-(1-tolylethenyl)phenyl]propane
B_{eff}	Blocking efficiency
BFPF	2,5-Bis[1-(2-furanyl)-1-methylethyl]-furan
2-Bu_3SnFu	2-*tert*-Butylstannylfuran
ε-CL	ε-Caprolactone
CMC	Critical micelle concentration
DFP	2,2-Difurylpropane
DiαMeSt-Cl	2-Chloro-2,4-diphenyl-4-methylpentane (adduct of α-methylstyrene dimer and hydrogen chloride)
DMFu	Difurylmethane
DPE	1,1-Diphenylethylene
DPn	Number-average degree of polymerization
DSC	Differential scanning calorimetry
DTBP	2,6-Di-*tert*-butylpyridine

DTE	1,1-Ditolylethylene
DVB	Divinylbenzene
FMF	2,5-Bis-(2-furylmethyl)furan
GPC	Gel permeation chromatography
GTP	Group transfer polymerization
HES	Hexaepoxy squalene
Hex	Hexane
IB	Isobutylene
IB-IB-Cl	Isobutylene-isobutylene end capped with chloride
IBVE	Isobutyl vinyl ether
K_{cd}	Equilibrium constant of capping/decapping equilibrium
k_c	Rate constant of capping
k_d	Rate constant of decapping
K_e	Apparent equilibrium constant
K_i	Equilibrium constant of ionization
MDDPE	1,3-Bis(1-phenylethenyl)benzene
MeChx	Methylcyclohexane
MeCl	Methyl chloride
αMeSt	α-Methylstyrene
MeVE	Methyl vinyl ether
MMA	Methyl methacrylate
M_n	Number-average molecular weight
N	Nucleophilicity parameter
NMR	Nuclear magnetic resonance (spectroscopy)
R_{cr}	Rate of crossover to a second monomer
R_p	Rate of homopolymerization of a second monomer
St	Styrene
pClαMeSt	*p*-Chloro-α-methylstyrene
pClSt	*p*-Chlorostyrene
PDDPE	1,4-Bis(1-phenylethenyl)benzene
PIB	Polyisobutylene
PIB-DPE$^+$	1,1-Diphenylethylene capped polyisobutylene carbocation
PIBVE	Poly(isobutyl vinyl ether)
pMeSt	*p*-Methylstyrene
PαMeSt	Poly(α-methylstyrene)
PMeVE	Poly(methyl vinyl ether)
PMMA	Poly(methyl methacrylate)
Poly(ε-CL-b-IB-b-ε-CL)	Poly(ε-caprolactone-*block*-isobutylene-*block*-ε-caprolactone) copolymer
Poly(IB-b-ε-CL)	Poly(isobutylene-*block*-ε-caprolactone) copolymer
Poly(IB-b-St)	Poly(isobutylene-*block*-styrene) copolymer
Poly(IB-b-IBVE)	Poly(isobutylene-*block*-isobutyl vinyl ether) copolymer
Poly(IB-b-αMeSt)	Poly(isobutylene-*block*-α-methylstyrene) copolymer
Poly(IB-b-MeVE)	Poly(isobutylene-*block*-methyl vinyl ether) copolymer
Poly(IB-b-MMA)	Poly(isobutylene-*block*-methyl methacrylate) copolymer

Poly(IB-b-tBMA)	Poly(isobutylene-*block*-*tert*-butyl methacrylate) copolymer
Poly(IB-b-PVL)	Poly(isobutylene-*block*-pivalolactone) copolymer
Poly(IB-s-IB′-s-MeVE)	Poly(isobutylene-*star*-isobutylene′-*star*-methyl vinyl ether) three-arm star-block copolymer
Poly(αMeSt-b-IB)	Poly(α-methylstyrene-*block*-isobutylene) copolymer
Poly(PVL-b-IB-b-PVL)	Poly(pivalolactone-*block*-isobutylene-*block*-pi valolactone) triblock copolymer
Poly(αMeSt-b-IB-b-αMeSt)	Poly(α-methylstyrene-*block*-isobutylene-*block*-α-methylstyrene) triblock copolymer
Poly(MMA-b-IB-b-MMA)	Poly(methyl methacrylate-*block*-isobutylene-*block*-methyl methacrylate) triblock copolymer
Poly(St-b-IB)	Poly(styrene-*block*-isobutylene) copolymer
Poly(pClαMeSt-b-IB)	Poly(*p*-chloro-α-methylstyrene-*block*-isobutylene) copolymer
Poly(pMeSt-b-IB-b-pMeSt)	Poly(*p*-methylstyrene-*block*-isobutylene-*block*-*p*-methylstyrene) triblock copolymer
Poly(St-b-IB-b-St)	Poly(styrene-*block*-isobutylene-*block*-styrene) triblock copolymer
PpClαMeSt	Poly(*p*-chloro-α-methylstyrene)
PPVL	Poly(pivalolactone)
PVL	Pivalolactone
PSt	Polystyrene
-St-IB-Cl	Dormant polymer end with styrene as the penultimate and isobutylene as the ultimate monomer unit
$t_{1/2}$	Half-life time of living chain ends
tBuDiCumCl	5-*tert*-Butyl-1,3-bis-(1-chloro-1-methylethyl)-benzene
T_g	Glass transition temperature
$[TiCl_4]_{free}$	Concentration of free and uncomplexed $TiCl_4$
T_m	Melting temperature
TMPCl	2-Chloro-2,4,4-trimethylpentane
TPE	Thermoplastic elastomer
σ	Tensile strength

1 Introduction

One of the most useful features of living polymerizations, which proceed in the absence of chain transfer to monomer and irreversible termination, is the ability to prepare block copolymers. Compared with living anionic polymerization the development of living cationic polymerization is rather recent. It was not until the mid-1980s that living cationic polymerization was reported first with vinyl ethers, and then with isobutylene (IB), and styrenic monomers [1–3]. Since then, great progress has been made in the field of living cationic polymerization for the synthesis of well-defined homopolymers and block copolymers with precisely controlled architectures. Particularly, PIB-based block copolymers have attracted considerable attention because of their unique properties such as UV and thermo-oxidative stability due to saturated backbone structure, high mechanical damping, high gas barrier property, biocompatibility, and biostability. Their commercial potential has recently been reviewed [4, 5].

In this chapter, synthetic methodologies for PIB-based block copolymers, as outlined in Scheme 1, are reviewed with examples developed mainly in our laboratory.

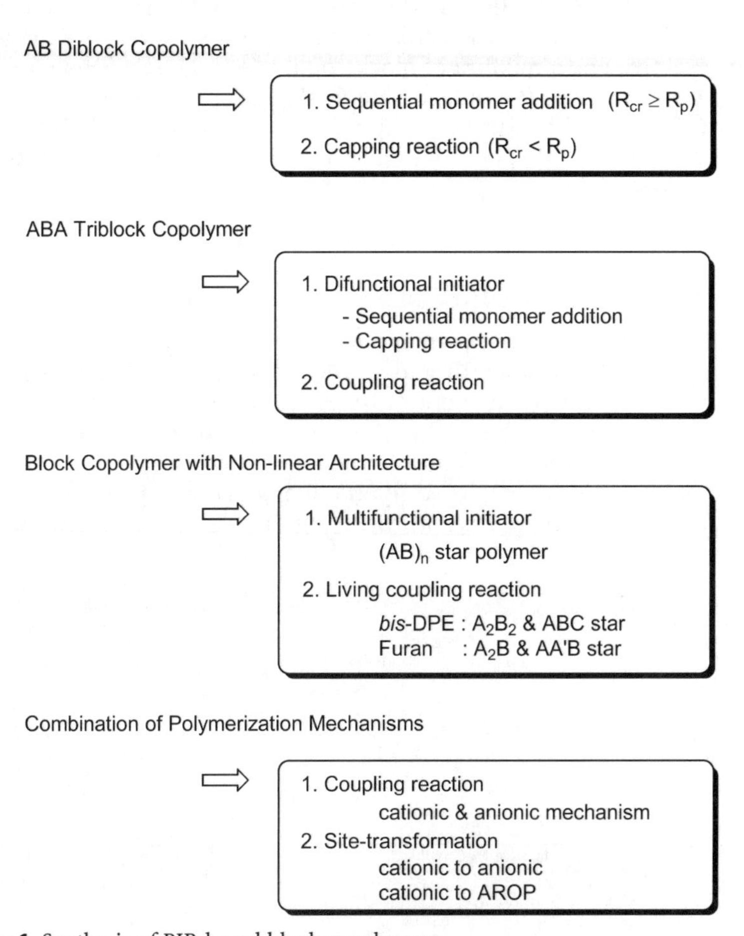

Scheme 1 Synthesis of PIB-based block copolymers

2
Linear Diblock Copolymers

Living cationic sequential block copolymerization is generally recognized as one of the simplest and most convenient methods of providing well-defined block copolymers with high structural integrity. The successful synthesis of block copolymers via sequential monomer addition relies on the rational selection of polymerization conditions such as Lewis acid, solvent, additives, temperature, etc., and on the selection of the appropriate order of monomer addition. For a successful living cationic sequential block copolymerization the rate of crossover to a second monomer (R_{cr}) must be i) faster than or ii) at least equal to that of the homopolymerization of a second monomer (R_p). In other words, efficient crossover could be achieved when the two monomers have similar reactivities or when crossover occurs from more reactive to less reactive monomer. When crossover is iii) from less reactive monomer to more reactive one a mixture of block copolymer and homopolymer is invariably formed due to the unfavorable R_{cr}/R_p ratio. The nucleophilicity parameter (N) reported by Mayr's group might be used as the relative scale of monomer reactivity [6]. The three cases mentioned above are discussed in detail as follows.

2.1
$R_{cr} \cong R_p$

When the reactivity of the two monomers is similar and steric factors are absent $R_{cr} \cong R_p$. For instance IB and styrene (St) possess similar reactivity, therefore, diblock copolymers poly(IB-*b*-St) [7] as well as the reverse order poly(St-*b*-IB) [8, 9] could be readily prepared via sequential monomer addition (Scheme 2). Moreover, identical reaction conditions (−80 °C and $TiCl_4$ as Lewis acid) could be employed for the living cationic polymerization of both monomers. However, whereas the living PIB chain ends are sufficiently

Cl + n ; $TiCl_4$, DTBP, -80 °C / Hex (or MeChx)/MeCl → m St → n m

Cl + n ; $TiCl_4$, DTBP, -80 °C / Hex (or MeChx)/MeCl → m IB → n-1 m-

Scheme 2 Synthesis of poly(IB-*b*-St) and poly(St-*b*-IB) diblock copolymers via sequential monomer addition

stable under monomer starved conditions, the living PSt chain ends undergo decomposition at close to ~100% conversion of St [10]. Some results suggest that this side reaction involves intramolecular hydride transfer followed by the disappearance of the tertiary cation in intramolecular alkylation. Therefore, IB must be added at $\leq$95% conversion of St in order to obtain poly(St-*b*-IB) diblock copolymers with negligible homoPSt contamination [10–12]. The presence of unreacted St monomer, however, complicates the block copolymerization of IB. After crossover the initial rate of IB polymerization is similar to that observed for homopolymerization initiated by a small molecule initiator. The first order plot, which is linear for homopolymerization, however, is curved downward for block copolymerization, indicating decreasing concentration and/or reactivity of active centers with time. This is attributed to the slow formation of -St-IB-Cl chain ends, which are much less reactive than -IB-IB-Cl [8, 9]. A similar effect of a γ-phenyl substituent has been observed by comparing the overall addition rate of 1,1-diphenylethylene and 2-phenylfuran to 2-chloro-2,4,4-trimethylpentane and 2-chloro-2,4-dimethyl-4-phenylpentane [13]. The overall reactivity of the latter compound was 30 times lower that that of the former one, due to much lower rate constant of ionization ascribed mainly to the negative inductive effect of the phenyl ring. In the case of -St-IB-Cl chain ends the reactivity may be further lowered by lower back strain relative to that of -IB-IB-Cl.

In the synthesis of the reverse sequence, poly(IB-*b*-St), it is important to add St after complete polymerization of IB. When St is added at less than 100% IB conversion the polymerization of St will be slow, which again is due to the formation and low reactivity of -St-IB-Cl chain ends. For instance when St is added after complete polymerization of IB, St polymerization is complete in 1 h. In contrast, when St is added at 94% IB conversion St conversion reaches only ~50% in 1 h at otherwise identical conditions.

2.2
$R_{cr}>R_p$

Rarely does one find references to living cationic sequential block copolymerization from a more reactive monomer to IB, for instance, from α-methylstyrene (αMeSt) to IB. Several reports on the living cationic polymerization of αMeSt have already been published using relatively mild Lewis acids such a $SnBr_4$, $TiCl_n(OR)_{4-n}$, or $SnCl_4$ as a coinitiator [14–17]. Since these Lewis acids are too weak to initiate the polymerization of less reactive IB, the addition of a stronger Lewis acid, e.g., $TiCl_4$ is necessary to obtain high molecular weight PIB after sequential monomer addition. With $SnBr_4$ or $TiCl_n(OR)_{4-n}$, however, ligand exchange takes place upon addition of $TiCl_4$, which results in mixed titanium halides that are too weak to initiate the polymerization of IB. Ligand exchange is absent with metal chlorides such as $SnCl_4$. However, with $SnCl_4$ the life-time of the propagating PαMeSt end is relatively short (half-life time of living PαMeSt chain ends ($t_{1/2}$)<2 min at −80 °C under conditions where close to complete conversion is reached in 2 min). The propagating PαMeSt end is relatively stable ($t_{1/2}\cong$3 h, ~100%

conversion in 4 min) using BCl_3 as a Lewis acid, which also induces living cationic polymerization of αMeSt in methylcyclohexane (MeChx)/methyl chloride (MeCl) (60/40 v/v) solvent mixture at −80 °C [18, 19]. Thus BCl_3 is suitable for the synthesis of poly(αMeSt-*b*-IB) diblock copolymer. The order of IB and $TiCl_4$ addition is critical, therefore, IB must be added first to avoid termination of living PαMeSt ends by intramolecular alkylation. Mechanistic studies on the early crossover step and propagation indicated that, upon addition of IB to the living PαMeSt solution, quantitative crossover took place followed by instantaneous termination (initiation without propagation) and the selective formation of PαMeSt-IB_1-Cl [20, 21]. However, the poly(αMeSt-*b*-IB) diblock copolymer obtained was contaminated by a small amount of PαMeSt homopolymer, apparently formed via intramolecular alkylation. This was surprising since initiation was very efficient using a model initiator, DiαMeSt-IB_1-Cl, and indanyl ring formation was absent. It was also shown that the blocking efficiency (B_{eff}) was linearly dependent on the concentration of IB, but this effect was substantially different for PαMeSt-IB_1-Cl and DiαMeSt-IB_1-Cl, which might be attributed to different local concentra-

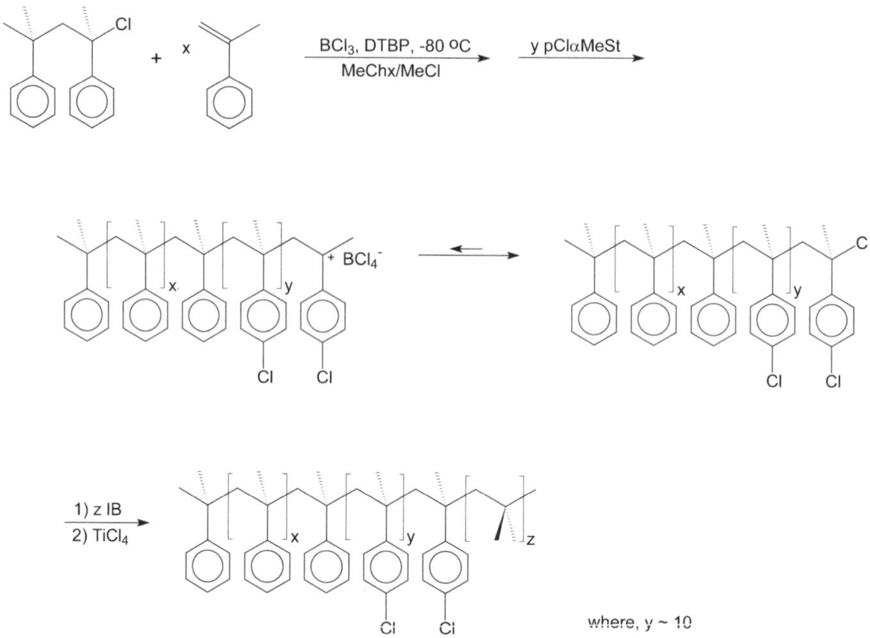

Scheme 3 Synthesis of poly(αMeSt-*b*-IB) copolymer by modifying the chain end of living PαMeSt, followed by sequential monomer addition

tions of IB, a poor solvent for PαMeSt. As presented in Scheme 3, intramolecular back-biting can be eliminated by modifying the PαMeSt chain end by adding *p*-chloro-α-methylstyrene (*p*ClαMeSt) (at [*p*ClαMeSt]/[living

PαMeSt chain end]$\geq$10) before the addition of IB. This is based on a recent finding that the living cationic polymerization of *p*ClαMeSt can be accomplished under conditions identical to those used for the synthesis of poly (αMeSt-*b*-IB) copolymer [22, 23]. Importantly, the living P*p*ClαMeSt chain end is very stable and there is no loss of livingness even after 5 h under monomer starved conditions. This is attributed to the reduced tendency of intramolecular alkylation due to the particularly large deactivating effect of the *p*-chloro substituent on the 2,5-positions of the aromatic ring.

Due to our interest in PIB-based block copolymers with a crystalline block segment the living cationic sequential block copolymerization of *p*ClαMeSt with IB was also studied. Sequential monomer addition has not been reported for this type of block copolymer due to limitation in the availability of monomers that undergo living cationic polymerization and give rise to crystalline polymers. Although cationic polymerization of αMeSt yields a highly syndiotactic polymer, it does not crystallize even when the syndiotactic content is higher than 90% [24]. Polymers of *p*-methyl- and *p*-chloro-substituted αMeSt have been reported to be semicrystalline with high T_g (140~175 °C) and T_m (210~220 °C) [25]. The living P*p*ClαMeSt end efficiently initiated the subsequent polymerization of IB, and poly(*p*ClαMeSt-*b*-IB) diblock copolymer was prepared via sequential monomer addition in conjunction with $TiCl_4$. On the basis of GPC UV traces of the starting P*p*ClαMeSt and the resulting poly(*p*ClαMeSt-*b*-IB) diblock copolymer, the B_{eff} was ~100% and homopolymer contamination was not detected. Crystallinity, however, was very low, estimated to be less than 5%.

2.3
$R_{cr}<R_p$ (Capping Reaction with Non-Homopolymerizable Monomers)

Sequential block copolymerization of IB with more reactive monomers such as αMeSt, *p*MeSt, isobutyl vinyl ether (IBVE), or methyl vinyl ether (MeVE) as a second monomer usually leads to a mixture of block copolymer and PIB homopolymer. As mentioned above, this is attributed to the unfavorable R_{cr}/R_p ratio. A special method for the synthesis of poly(IB-*b*-MeVE) copolymers was reported by Kennedy et al. [26, 27] using the TMPCl/$TiCl_4$ initiating system in the presence of Bu_4NCl in Hex/MeCl or Hex/CH_2Cl_2 at −80 °C, followed by sequential addition of MeVE. Although MeVE is much more reactive than IB, the rate of propagation of MeVE decreases with conversion and the polymerization stops short of completion when all Lewis acid is complexed with the polymer, suggesting strong complexation between $TiCl_4$ and the ether oxygen. The polymerization can be restarted by a further addition of $TiCl_4$; however, complete polymerization of MeVE can only be achieved in the presence of excess $TiCl_4$ relative to the monomer. Dealcoholation, a side reaction that usually accompanies the polymerization of alkyl vinyl ethers in the presence of strong Lewis acids, was observed; however, it could be prevented by the use of tetrabutylammonium chloride (Bu_4NCl). The synthesis of poly(IB-*b*-IBVE) copolymer was also attempted by sequential monomer addition using the TMPCl/$TiCl_4$ initiating system in the presence of $AcNMe_2$ [28]. The degradation of PIBVE block by dealcoholation was prevented using low temperature (−100 °C).

To overcome the difficulty in the crossover step a general methodology has been developed in our laboratory for the synthesis of block copolymers when the second monomer is more reactive than the first one. It involves the intermediate capping reaction with non-(homo)polymerizable monomers such as i) 1,1-diphenylethylene (DPE) and its derivatives and ii) 2-substituted furans.

2.3.1
Synthesis Using 1,1-Diarylethylenes

As shown in Scheme 4 [29], this process involves the capping reaction of living PIB with DPE or 1,1-ditolylethylene (DTE), followed by tuning of the

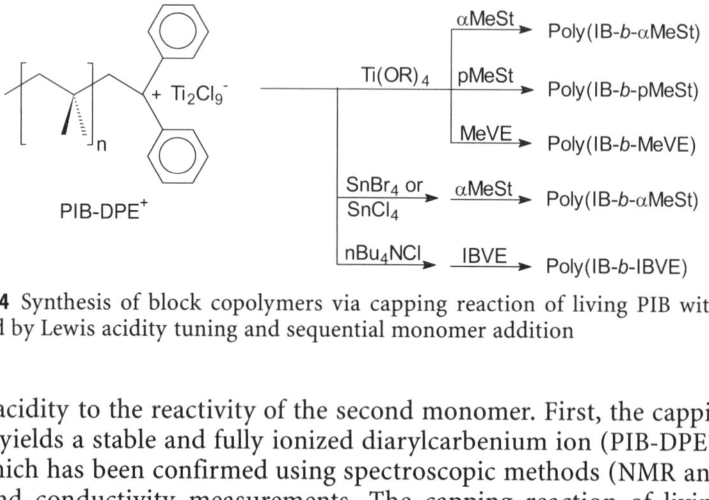

Scheme 4 Synthesis of block copolymers via capping reaction of living PIB with DPE, followed by Lewis acidity tuning and sequential monomer addition

Lewis acidity to the reactivity of the second monomer. First, the capping reaction yields a stable and fully ionized diarylcarbenium ion (PIB-DPE$^+$) [30, 31], which has been confirmed using spectroscopic methods (NMR and UV/Vis) and conductivity measurements. The capping reaction of living PIB with 1,1-diarylethylenes is an equilibrium reaction, which can be shifted toward completion with decreasing temperature, or with increasing Lewis acidity, solvent polarity, electron-donating ability of *p*-substituents, or concentration of reactants. The corresponding thermodynamic ($K_e=K_iK_{cd}$) and kinetic constants (k_cK_i and k_d) of the capping/decapping reaction have been determined [32–34]. The capping reaction is also applicable to living polystyrene [35, 36]; however, capping is irreversible, i.e., decapping of the chain ends does not occur. The purpose of the Lewis acidity tuning, following the capping reaction, is to generate more nucleophilic counterions, which ensure a high R_{cr}/R_p ratio as well as the living polymerization of a second monomer. This has been carried out using three different methods: (i) by the addition of titanium(IV) alkoxides ($Ti(OR)_4$), (ii) by the substitution of a strong Lewis acid with a weaker one, or (iii) by the addition of n-Bu_4NCl.

The first and best method has been successfully employed in the block copolymerization of IB with αMeSt [37], *p*MeSt [38], or MeVE [39, 40].

The substitution of $TiCl_4$ with a weaker Lewis acid ($SnBr_4$ or $SnCl_4$) has also been proven to be an efficient strategy in the synthesis of poly(IB-*b*-αMeSt) diblock copolymer [16, 17]. The disadvantage of this technique is that $TiCl_4$ needs to be deactivated by a suitable Lewis base before the addition of αMeSt and $SnBr_4$ or $SnCl_4$. When $SnCl_4$ was employed, it was necessary to keep $[SnCl_4]$ equal to or below 0.5×[chain end] to increase the ratio of R_{cr}/R_p. Mechanistic studies indicated that, when $[SnCl_4]$~0.5×[chain end], a double charged counterion, $SnCl_6^{2-}$, was involved during the crossover reaction and converted to a single charged counterion, $SnCl_5^-$, during the polymerization of αMeSt.

The block copolymerization of IB with IBVE was achieved by Lewis acidity tuning using *n*-Bu_4NCl [41, 42]. The addition of nBu_4NCl reduces the concentration of free and uncomplexed $TiCl_4$ ($[TiCl_4]_{free}$), and mechanistic studies indicated that, when $[TiCl_4]_{free}$<[chain end], the dimeric counterion, $Ti_2Cl_9^-$, was converted to a more nucleophilic monomeric $TiCl_5^-$ counterion suitable for the living polymerization of IBVE.

2.3.2
Synthesis Using Furan Derivatives

Successful block copolymerization of IB with MeVE has also been accomplished using 2-alkylfurans as a new class of non-(homo)polymerizable monomers. We have discovered that certain O, S, or N containing heterocyclic diolefins such as 2-substituted furan, thiophene, pyrrole, etc., add to living PIB quantitatively without homopolymerization. With less reactive thiophene or pyrrole even the unsubstituted compounds can be used. The resulting cationic charge is retained at the polymer chain end and may provide an efficient initiating site for certain monomers possessing high reactivity [43]. In particular, the reaction of living PIB with 2-methyl-, 2-*tert*-butyl-, and 2-phenylfuran has been studied in detail and utilized in living cationic block copolymerization. Mechanistic studies reveal that the addition occurs exclusively at the C-5 position on the furan ring and a stable tertiary allylic cation is generated at the C-2 position, which is further stabilized by the neighboring electron rich oxygen atom as shown in Scheme 5 [44]. The formation of the stable allylic cation was further confirmed by UV-Vis spectroscopy for 2-phenylfuran and from trapping experiment of the furanyl cations with tributyltin hydride, which resulted in PIB with 2,5-dihydrofuran functionality in quantitative yield. Interestingly, quenching with methanol produced the 2,5-substituted furan in quantitative yield, most probably due to the intermediate formation of an acetal, which eliminated methanol.

One of the promising features of the capping reaction of living PIB with furan derivatives is that rapid and quantitative mono-addition to living PIB has been observed not only in conjunction with $TiCl_4$ as Lewis acid in Hex/CH_2Cl_2 (or MeCl) (60/40 v/v) at −80 °C but also with BCl_3 in MeCl even at −40 °C. Compared to DPE and its derivatives, the resulting tertiary allylic

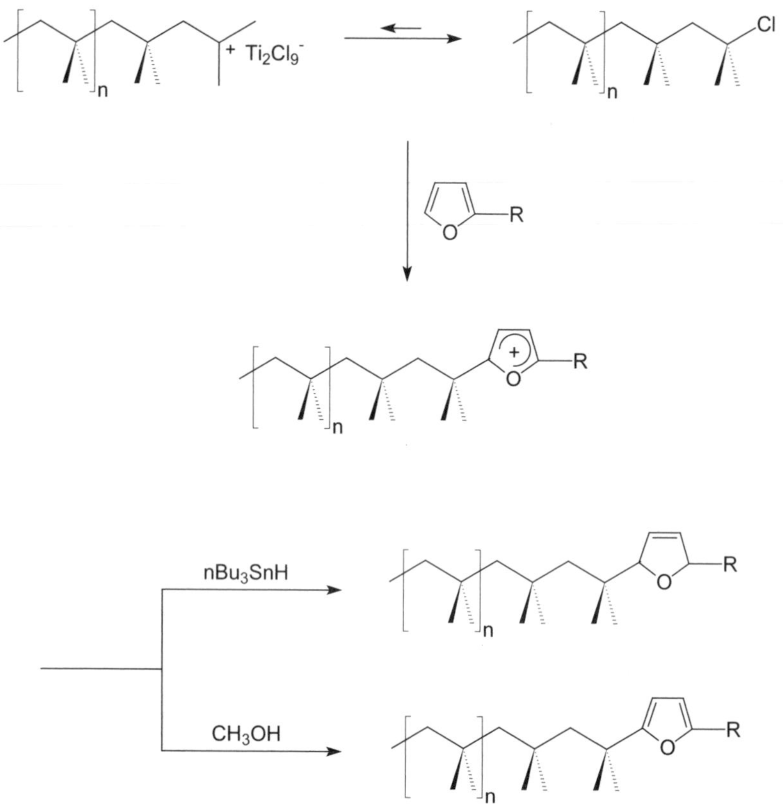

Scheme 5 Capping reaction of living PIB with 2-alkylfurans

cation is more stable and retroaddition (decapping) does not take place even at –20 °C [45]. However, a slow decomposition of the furanyl cation by proton elimination is observed at –20 °C in conjunction with $TiCl_4$. Therefore, it appears that 2-alkylfurans are more suitable capping agents when the subsequent functionalization or block copolymerization are carried out at elevated temperature.

Block copolymerization of IB with MeVE was carried out using 2-methylfuran or 2-*tert*-butylfuran as a capping agent [44, 46]. The crossover efficiency of ~66% was obtained using 2-*tert*-butylfuran, slightly higher (~75%) when 2-methylfuran was employed as a capping agent under similar condition. Structural analysis of the products showed that the polymerization of MeVE was initiated at the C-4 position. This can be rationalized not only by the larger steric hindrance at the C-2 position but by the formation of a more reactive (i.e., less stable) cation at the C-4 position.

3
Linear Triblock Copolymers

3.1
Synthesis Using Difunctional Initiators

Since soluble multifunctional initiators are more readily available in cationic polymerization than in the anionic counterpart, ABA type linear triblock copolymers have been almost exclusively prepared using difunctional initiation followed by sequential monomer addition. The preparation and properties of ABA type block copolymer thermoplastic elastomers (TPEs), where the middle segment is PIB, have been reviewed recently [47].

The synthesis of poly(St-*b*-IB-*b*-St) triblock copolymer has been accomplished by many research groups [48–54]. The synthesis invariably involved sequential monomer addition using a difunctional initiator in conjunction with $TiCl_4$ in a moderately polar solvent mixture at low (−70 to −90 °C) temperatures. As already mentioned at the synthesis of poly(IB-*b*-St) it is important to add St at ~100% IB conversion. The selection of the solvent is also critical, because coupled product that forms in intermolecular alkylation during St polymerization can not be avoided when the solvent is a poor solvent (e.g., Hex/MeCl 60/40 v/v) for PSt [55]. The formation of coupled product is slower in *n*-BuCl or in MeChx/MeCl 60/40 v/v solvent mixture. However, to obtain block copolymers essentially free of coupled product it is necessary to stop the polymerization of St before completion. Studies on the relationship of macromolecular architecture/mechanical properties indicate that the tensile strength is controlled by the molecular weight of the PSt segment and is independent of the PIB middle block length. This can be attributed to the morphological development; phase-separation starts when the $\bar{M}_n$ of the PSt segment reaches ~5000, and it is complete at $\bar{M}_n$~15,000. Further increase in the PSt block length therefore does not increase the tensile strength. Representative triblocks exhibited 23–25 MPa of tensile strength [52, 56], similar to that of commercially available styrenic TPEs obtained by anionic polymerization. The two step sequential monomer addition method has also been employed to obtain poly(*p*-chlorostyrene-*b*-IB-*b*-*p*-chlorostyrene) [57], poly(indene-*b*-IB-*b*-indene) [58], poly(*p*-*tert*-butylstyrene-*b*-IB-*b*-*p*-*tert*-butylstyrene) [59], poly((indene-*co*-*p*-methylstyrene)-*b*-IB-*b*-(indene-*co*-*p*-methylstyrene)) [60], and poly(*p*MeSt-*b*-IB-*b*-*p*MeSt) [61] copolymers.

When the crossover from the living PIB chain ends is slower than propagation of a second monomer, e.g., αMeSt and *p*MeSt, the final product is invariably a mixture of triblock and diblock copolymers and possibly homoPIB, which results in low tensile strength and low elongation (4.6 MPa at 260% elongation) [62]. This slow crossover can be circumvented by the synthetic strategy shown above utilizing an intermediate capping reaction of the living PIB with DPE followed by moderating the Lewis acidity before the addition of the second monomer. This method has been successfully employed for the synthesis of poly(αMeSt-*b*-IB-*b*-αMeSt) [63] and poly(*p*MeSt-

b-IB-*b*-*p*MeSt) [64]. Tensile strength of these TPEs as well as triblock copolymers reported above was similar to that obtained with poly(St-*b*-IB-*b*-St) and virtually identical to that of vulcanized butyl rubber, indicating failure in the elastomeric domain.

3.2 Synthesis Using Coupling Agents

Although the synthetic strategy using non-(homo)polymerizable monomers presented above has been shown to be highly effective for the synthesis of a variety of di- or triblock copolymers, ABA type linear triblock copolymers can be also prepared by coupling of living diblock copolymers. This is a general method in living anionic polymerization. Several coupling agents for living poly(vinyl ethers) and PαMeSt have been reported in cationic polymerization [65–67], but quantitative coupling was limited to the living polymers with low molecular weight (DPn~10). We have recently extended the concept of capping reaction to coupling reaction of living PIB using bis-DPE and bis-furanyl compounds [68, 69].

3.2.1 *Synthesis Using Bis-DPE Derivatives*

Synthetic utilization of non-(homo)polymerizable diolefins has been first shown for the coupling reaction of living PIB [69, 70]. Using 2,2-bis[4-(1-phenylethenyl)phenyl]propane (BDPEP) or 2,2-bis[4-(1-tolylethenyl)phenyl]propane (BDTEP) as a coupling agent (Scheme 6), a rapid and quantita-

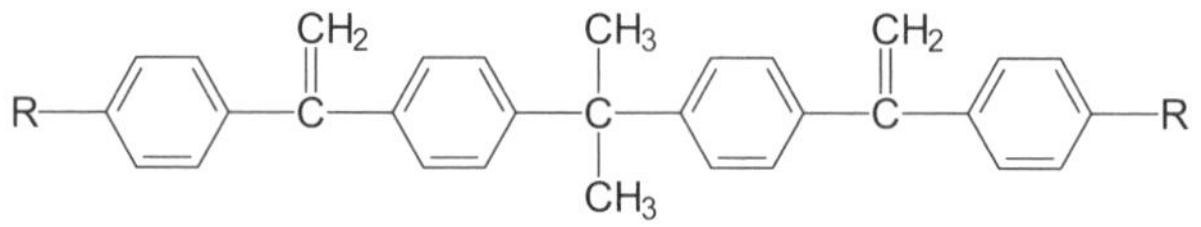

Scheme 6 Structures of bis-DPE compounds; R=H (BDPEP), R=CH_3 (BDTEP)

tive coupling reaction of living PIB (Scheme 7) was achieved, independently of the molecular weight of PIB. Kinetic studies indicated that coupling reaction of living PIB by bis-DPE compounds was a consecutive reaction where the second addition was much faster than the first one. As a result, high coupling efficiency was also observed, even when excess BDPEP was used. This coupling agent is therefore best suited for the synthesis of ABA triblock copolymers by coupling of living AB diblock copolymers, and has been employed to obtain poly(St-*b*-IB-*b*-St) [9] and poly(αMeSt-*b*-IB-*b*-αMeSt) [22] triblock copolymers. For the synthesis of poly(St-*b*-IB-*b*-St) triblock copolymers, however, the two step monomer addition method is superior. We recall that due to decomposition of the living PSt chain ends at close to complete St conversion, IB must be added at $\leq$95% St conversion to obtain living

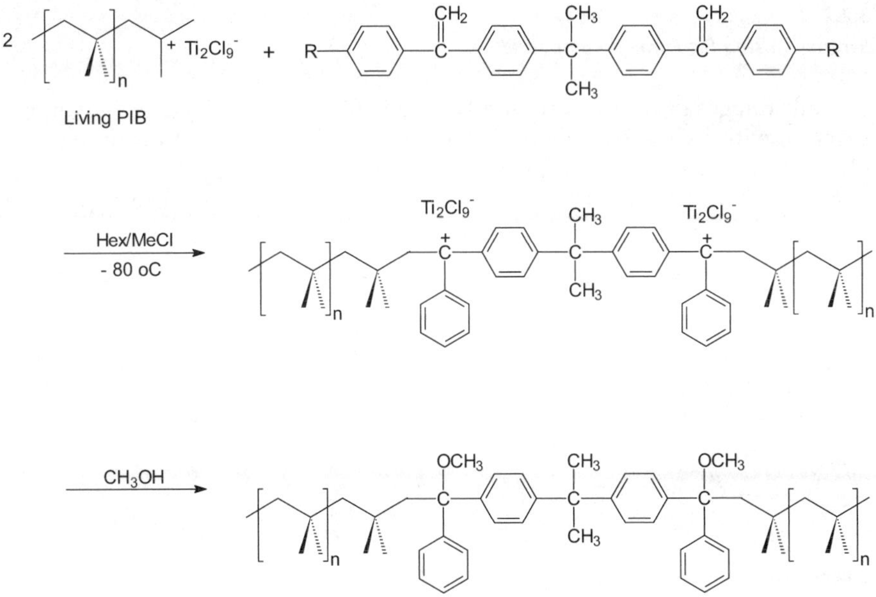

Scheme 7 Coupling reaction of living PIB by bis-DPE compound

poly(St-*b*-IB) diblocks with negligible PSt homopolymer contamination. The residual unreacted St in the subsequent polymerization of IB, however, gives rise to a relatively high concentration of unreactive -St-IB-Cl chain ends, which causes coupling of living poly(St-*b*-IB) diblocks to be very slow and incomplete even after 50 h. Due to diblock contaminants in the final product the mechanical properties were satisfactory (σ=16–20 MPa) but inferior to the best triblocks made by a difunctional initiation, followed by sequential monomer addition.

In the case of poly(αMeSt-*b*-IB-*b*-αMeSt) triblock copolymer, coupling could be the method of choice, since crossover from living PIB to αMeSt is unfavorable. Poly(αMeSt-*b*-IB) diblock copolymers were synthesized by living sequential cationic polymerization in MeChx/MeCl solvent mixtures at −80 °C using BCl_3 for the polymerization of αMeSt and $TiCl_4$ as the coinitiator for the polymerization of IB. By modifying the living PαMeSt chain end with a short segment of *p*ClαMeSt before adding IB, the crossover efficiency of ~90% was achieved, as shown in Sect. 2.2. More importantly, subsequent in situ coupling of the living block copolymers with BDPEP was rapid and nearly quantitative [22].

3.2.2
Synthesis Using Bis-Furanyl Compounds

The advantages of furan compounds over DPE and its derivatives in capping reaction with living PIB mentioned above have been utilized for living coupling reaction using bis-furanyl compounds as coupling agents.

Several bis-furanyl compounds shown in Scheme 8 were investigated for

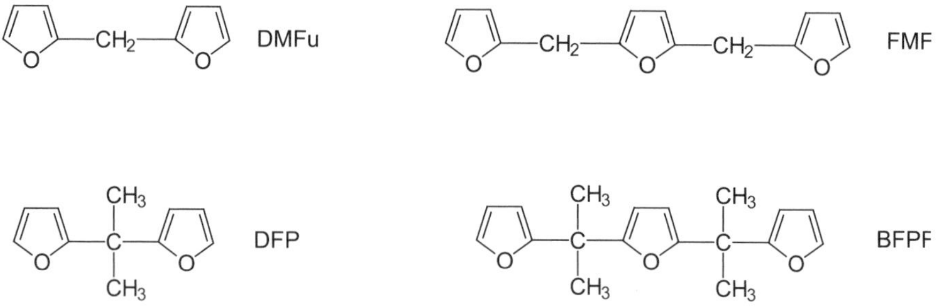

DMFu FMF

DFP BFPF

Scheme 8 Structures of bis-furanyl compounds

coupling living PIB. Using bis-furanyl compounds with a short spacer group such as difurylmethane (DMF) and 2,2-difurylpropane (DFP) coupling was less than quantitative (35 and ~50% respectively) suggesting that the reactivity of the second furan ring was decreased significantly upon monoaddition. A relatively high coupling efficiency was obtained with 2,5-bis-(2-furylmethyl)furan (FMF) as coupling agents (~85%); however, the coupled product exhibited a strong orange color, indicating the presence of well-known side reaction, i.e., hydride abstraction at α-position to the ring [71]. This side reaction was effectively avoided using 2,5-bis[1-(2-furanyl)-1-methylethyl]-furan (BFPF), and quantitative coupling of living PIB was observed as presented in Scheme 9.

More importantly, the coupling reaction of living PIB with BFPF could also be achieved with the BCl_3/MeCl/−40 °C system for which coupling using bis-DPE compounds as coupling agents was not applicable.

4
Block Copolymers with Non-Linear Architecture

Cationic synthesis of block copolymers with non-linear architectures has been reviewed recently [72]. These block copolymers have served as model materials for systematic studies on architecture/property relationships of macromolecules. $(AB)_n$ type star-block copolymers, where n represents the number of arms, have been prepared by the living cationic polymerization using three different methods: (i) via multifunctional initiators, (ii) via multifunctional coupling agents, and (iii) via linking agents.

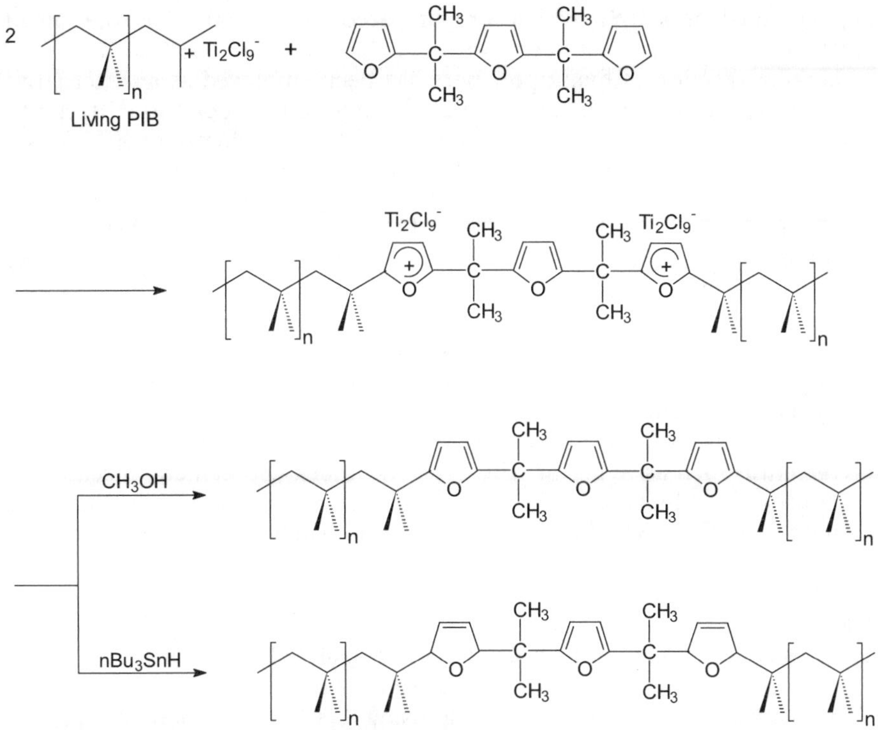

Scheme 9 Coupling reaction of living PIB with BFPF

The synthesis using multifunctional initiators has been the most versatile method due to the affluence of well-defined soluble multifunctional initiators for a variety of monomers. Using trifunctional initiators many groups have prepared three arm star-block copolymers such as poly(IBVE-*b*-2-hydroxyethyl vinyl ether)$_3$ [73], poly(IB-*b*-St)$_3$ [74, 75], and poly(IB-*b*-*p*MeSt)$_3$ [76] star-block copolymers. The synthesis of eight arm poly(IB-*b*-St)$_8$ star-block copolymers was reported recently [77]. An octafunctional calix[8]arene-based initiator was used to initiate the living cationic polymerization of IB to desirable molecular weight followed by sequential addition of St to obtain the star-block copolymer. Star-block TPEs containing 17–32% PSt showed high strength and elongation (up to ~26 MPa and >500%) and relatively low melt viscosity. The rheological properties of these star-blocks were superior to those of poly(St-*b*-IB-*b*-St) linear triblock copolymers. The synthesis of poly(IB-*b*-*p*-chlorostyrene (*p*ClSt))$_8$ star-block copolymer was accomplished using the previously mentioned method [78]. The star-block copolymer exhibited high tensile strengths (22–27 MPa) and elongations (~500%). Recently, multi-arm star-block copolymers of poly(IB-*b*-St) [79] and poly(IB-*b*-*p*-*tert*-butylstyrene) [80] copolymers were synthesized by living cationic polymerization using a hexafunctional initiator, hexaepoxy squalene (HES), which was prepared by a simple epoxidation of squalene.

Linking reaction of living polymers has been employed as an alternative way to prepare star-block copolymers. The synthesis of poly(St-*b*-IB) multiarm star-block copolymers was reported using divinylbenzene (DVB) as a linking agent [81]. The synthesis and mechanical properties of star-block copolymers consisting of 5–21 poly(St-*b*-IB) arms emanating from cyclosiloxane cores have been published [82]. The synthesis involved the sequential living cationic block copolymerization of St and IB, followed by quantitative allylic chain end-functionalization of the living poly(St-*b*-IB), and finally linking of these prearms with SiH-containing cyclosiloxanes (2,4,6,8,10,12-hexamethylcyclohexasiloxane) by hydrosilation. Star-block copolymers of poly(indene(Ind)-*b*-IB) have been prepared using the previously mentioned method [83].

The synthesis of hetero-arm star-block copolymers with well-controlled architecture such as A_nB_n- or ABC-type star-block copolymers, recently accomplished by a living anionic process utilizing a novel concept of the living coupling reaction [84], however, has been a challenge by a living cationic process.

4.1 Synthesis of Hetero-Arm Star-Block Copolymers Using Bis-DPE Derivatives

The coupling reaction of living PIB using bis-DPE compounds such as BDPEP or BDTEP as coupling agents has been presented above [69]. It was also conceived that bis-DPE compounds could be useful as "living" coupling agents for living PIB. According to the definition, a living coupling agent must react quantitatively with the living chain ends and the coupled product must retain the living active centers stoichiometrically. For block copolymer synthesis the living coupled product should be able to reinitiate the second monomer rapidly and stoichiometrically. We demonstrated that living PIB reacted quantitatively with these coupling agents to yield stoichiometric amounts of bis(diarylalkylcarbenium) ions. Since diarylalkylcarbenium ions have been shown above to be successful for the controlled initiation of reactive monomers such as *p*MeSt, αMeSt, IBVE, and MeVE, it is apparent that bis-DPE compounds are highly qualified as living coupling agents, satisfying all the criteria mentioned above.

As a proof of the concept, an amphiphilic A_2B_2 star-block copolymer (A=PIB and B=PMeVE) has been prepared by the living coupling reaction of living PIB followed by the chain-ramification polymerization of MeVE at the junction of the living coupled PIB as shown in Scheme 10 [85]. Architecture/property studies were carried out by comparing the micellar properties of star-block copolymer ($(IB_{45})_2$-*b*-$(MeVE_{170})_2$) and the corresponding linear diblock analogues with same total molecular weight and composition (IB_{90}-*b*-$MeVE_{340}$) or with same segmental lengths (IB_{45}-*b*-$MeVE_{170}$). A_2B_2 star-block copolymers exhibited an order of magnitude higher critical micelle concentration (CMC) in aqueous solution, compared to that of the corre-

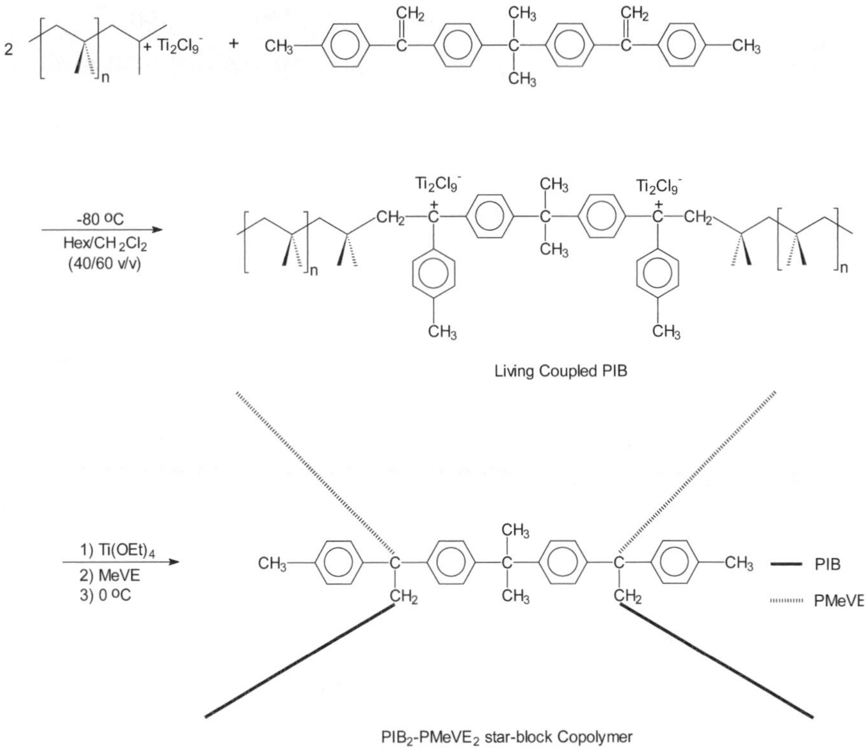

Living Coupled PIB

PIB$_2$-PMeVE$_2$ star-block Copolymer

Scheme 10 Living coupling reaction of living PIB with BDTEP and chain ramification reaction of MeVE for the synthesis of A_2B_2 star-block copolymer

sponding linear diblock copolymers, indicating that the micellar properties of star-block and linear diblock copolymers are strongly dependent on the block architecture. Interestingly the hydrodynamic radius was also much larger for ((IB$_{45}$)$_2$-*b*-(MeVE$_{170}$)$_2$, R_h=88 nm) than the corresponding (IB$_{45}$-*b*-MeVE$_{170}$, R_h=21 nm), suggesting highly stretched chains.

During the study of the living coupling reaction of living PIB using bis-DPE compounds, "double" diphenylethylenes such as 1,3-bis (1-phenylethenyl)benzene (MDDPE) and 1,4-bis(1-phenylethenyl)benzene (PDDPE), which are living anionic coupling agents, have also been examined. Coupling, however, is not observed, indicating that the second double bond is far less reactive than the first one. Taking advantage of the quantitative monoaddition, a facile route for the synthesis of PIB-DPE macromonomers has been developed [86]. On the basis of spectroscopic as well as chromatographic results, PDDPE is a better candidate than MDDPE for the synthesis of the macromonomers. When PDDPE was employed, the formation of the coupled product was not detected, even when 2 equivalent of PDDPE was employed. Moreover PDDPE is ~2.5 times more reactive than MDDPE toward living PIB. Using these types of macromonomers, the synthesis of

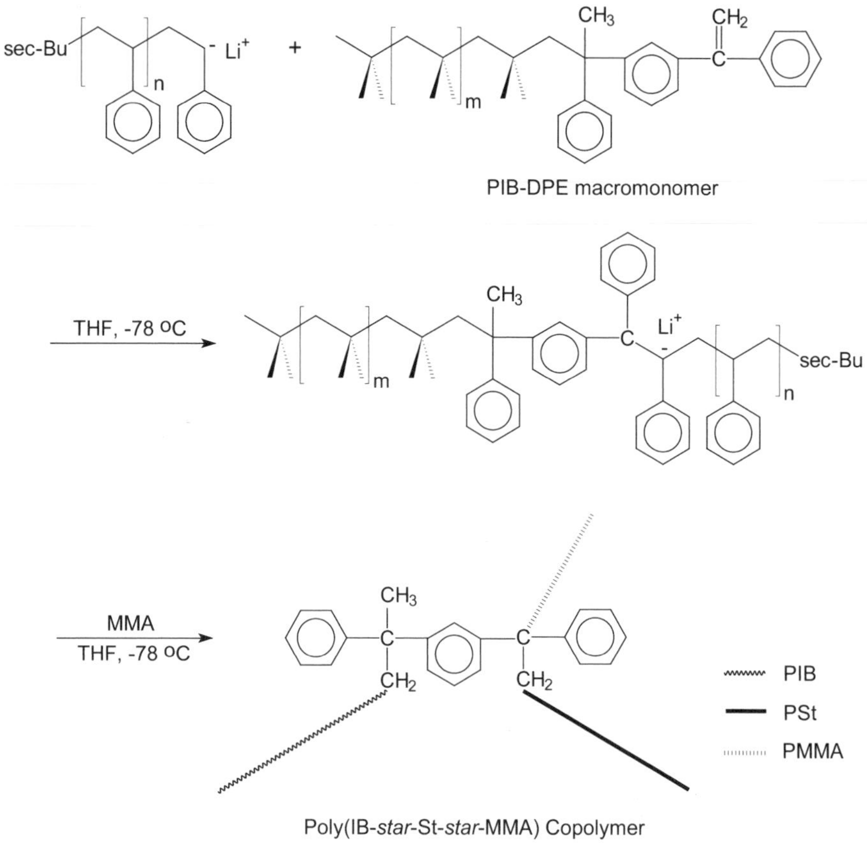

Scheme 11 Synthesis of ABC star-block copolymer

ABC-type star-block copolymers is currently under investigation in our laboratory as shown in Scheme 11.

4.2 Synthesis of AA′B, ABB′, and ABC Asymmetric Star-Block Copolymers Using Furan Derivatives

While the concept of coupling with ω-furan functionalized PIB as a polymeric coupling agent has been utilized to obtain AB-type block copolymers, it is also apparent that ω-furan functionalized polymers can be used as living coupling polymeric precursor for the synthesis of AA′B-, ABB′- and ABC-type three-arm star-block copolymers, where A (and A′) and B (and B′) represent PIB and PMeVE segments, respectively, with different molecular weights, and C represents chemically different block segment such a PSt.

To illustrate this concept, the strategy for the synthesis of AA′B-type star-block copolymers, where A=PIB(1), A′=PIB(2), and B=PMeVE, is shown in

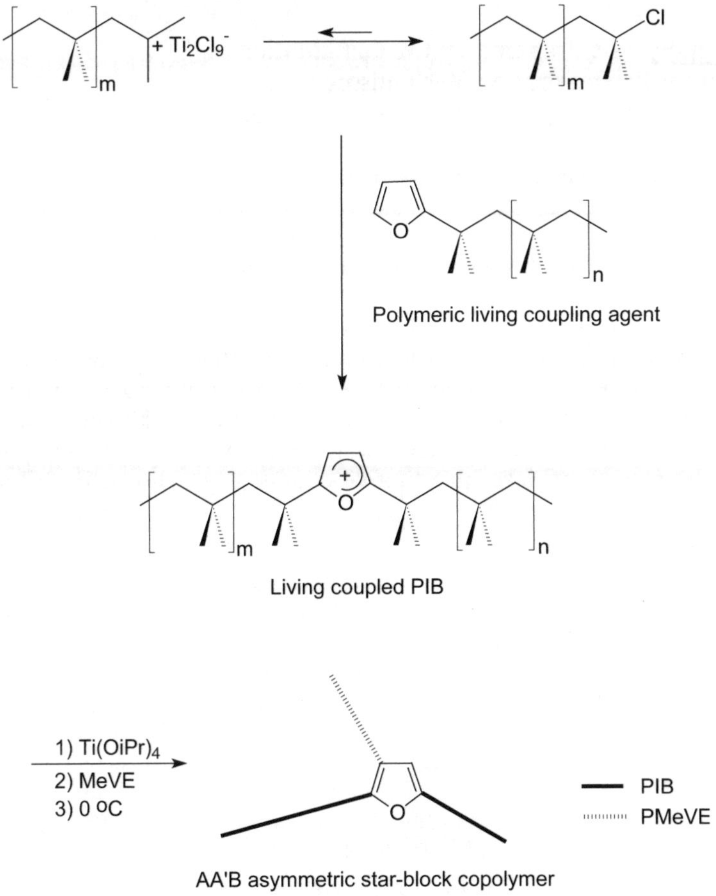

+ $Ti_2Cl_9^-$

Cl

Polymeric living coupling agent

Living coupled PIB

1) $Ti(OiPr)_4$
2) MeVE
3) 0 °C

PIB
PMeVE

AA'B asymmetric star-block copolymer

Scheme 12 Synthesis of AA′B asymmetric star-block copolymer

Scheme 12. First, quantitative addition of ω-furan functionalized PIB (A′), obtained from a simple reaction between living PIB and 2-Bu_3SnFu, to living PIB (A) could be achieved in Hex/CH_2Cl_2 (40/60 v/v) at −80 °C in conjunction with $TiCl_4$. The resulting living coupled PIB-Fu^+-PIB′ was successfully employed for the subsequent chain ramification polymerization of MeVE. This technique has its unique ability to control A and A′ block length independently. Pure poly(IB-*s*-IB′-*s*-MeVE) three-arm star-block copolymer was obtained upon purification of the crude product by column chromatography [87].

5 Block Copolymers Prepared by the Combination of Different Polymerization Mechanisms

5.1 Combination of Cationic and Anionic Polymerization

The combination of living cationic and anionic techniques provides a unique approach to block copolymers not available by a single method. Site transformation and coupling of two homopolymers are convenient and efficient ways to prepare well-defined block copolymers.

Block copolymers of IB and MMA monomers that are polymerizable only by different mechanisms can be prepared by several methods. The prerequisite for the coupling reaction is that the reactivities of the end groups have to be matched and a good solvent has to be found for both homopolymers and copolymer to achieve quantitative coupling. As shown in Scheme 13,

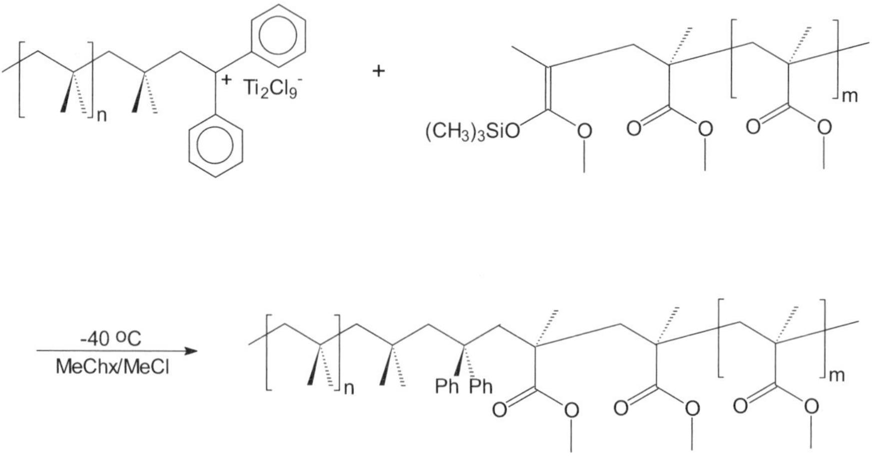

Scheme 13 Synthesis of poly(IB-*b*-MMA) block copolymers by coupling reaction

poly(IB-*b*-MMA) block copolymers were synthesized by coupling reaction of two corresponding living homopolymers, obtained by living cationic and group transfer polymerization (GTP), respectively [88]. It was found that after reaching a maximum coupling efficiency of ~80% slow decoupling took place with increasing reaction time. This competitive decoupling reaction was more evident at higher temperatures and high [$TiCl_4$], and proposed to involve the complexation of $TiCl_4$ with the carbonyl oxygen adjacent to the coupling site followed by anion formation and irreversible chain scission. This side reaction was negligible when the reaction time was less than 4 h at −40 °C and [$TiCl_4$]/[$PIBDPE^+Ti_2Cl_9^-$]=4.

The synthesis of poly(MMA-*b*-IB-*b*-MMA) triblock copolymers has also been reported using the site-transformation method, where α,ω-dilithiated PIB was used as the macroinitiator [89]. The site-transformation technique provides a useful alternative for the synthesis of block copolymers consisting of two monomers that are polymerized only by two different mechanisms. In this method, the propagating active center is transformed to a different kind of active center and a second monomer is subsequently polymerized by a mechanism different from the preceding one. The key process in this method is the precocious control of α or ω-end functionality, capable of initiating the second monomer. Recently a novel site-transformation reaction, the quantitative metalation of DPE-capped PIB carrying methoxy or olefin functional groups, has been reported [90]. This method has been successfully employed in the synthesis of poly(IB-*b*-*t*BMA) diblock and poly (MMA-*b*-IB-*b*-MMA) triblock copolymers [91].

5.2 Combination of Living Cationic and Anionic Ring-Opening Polymerization for the Synthesis of Semicrystalline Block Copolymers

Block copolymers containing crystallizable blocks have been studied not only as alternative TPEs with improved properties but also as novel nanostructured materials with much more intricate architectures compared to those produced by the simple amorphous blocks. Since the interplay of crystallization and microphase segregation of crystalline/amorphous block copolymers greatly influences the final equilibrium ordered states, and results in a diverse morphological complexity, there has been a continued high level of interest in the synthesis and characterization of these materials.

Due to the lack of vinyl monomers giving rise to crystalline segment by cationic polymerization, amorphous/crystalline block copolymers have not been prepared by living cationic sequential block copolymerization. Although site-transformation has been utilized extensively for the synthesis of block copolymers, only a few PIB/crystalline block copolymers such as poly(*L*-lactide-*b*-IB-*b*-*L*-lactide) [92], poly(IB-*b*-ε-caprolactone(ε-CL)) [93] diblock and poly(ε-CL-*b*-IB-*b*-ε-CL) [94] triblock copolymers with relatively short PIB block segment ($\bar{M}_n \leq 10{,}000$ g/mol) were reported. This is most likely due to difficulties in quantitative end-functionalization of high molecular weight PIB.

We recently investigated a different route for the synthesis of poly(IB-*b*-ε-CL) diblock and poly(ε-CL-*b*-IB-*b*- ε-CL) triblock copolymers by site-transformation of living cationic polymerization of IB to cationic ring-opening polymerization of ε-CL via the "activated monomer mechanism" [95].

The synthesis of poly(IB-*b*-pivalolactone (PVL)) diblock copolymers was also recently accomplished by site-transformation of living cationic polymerization of IB to AROP of PVL, as shown in Scheme 14 [96, 97]. First, PIB with ω-carboxylate potassium salt was prepared by capping living PIB with DPE followed by quenching with 1-methoxy-1-trimethylsiloxy-propene (MTSP), and hydrolysis of ω-methoxycarbonyl end groups. The ω-carboxyl-

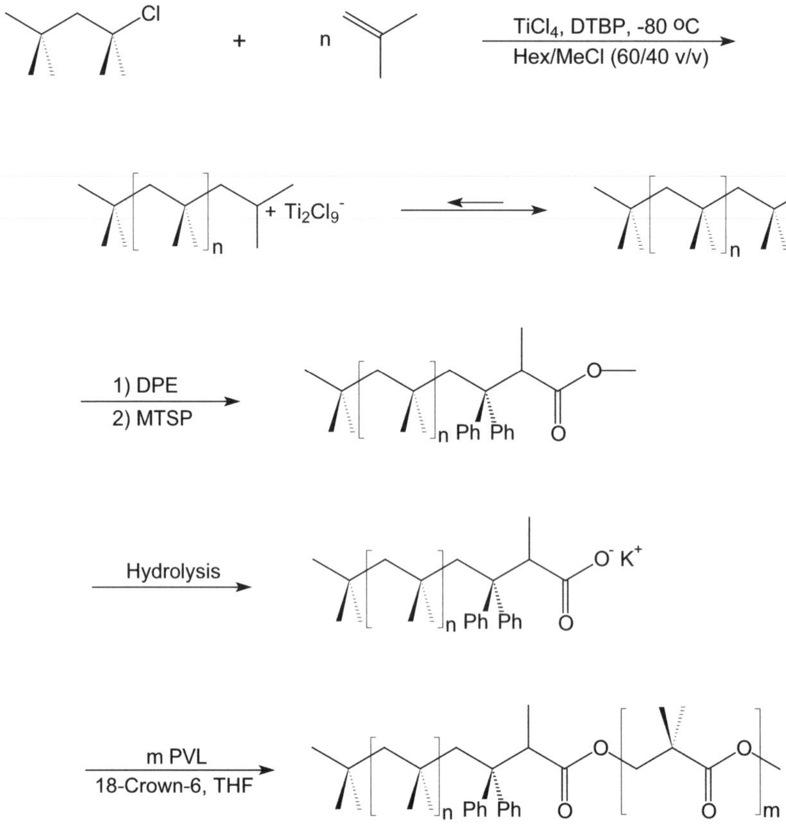

$TiCl_4$, DTBP, -80 °C
Hex/MeCl (60/40 v/v)

+ $Ti_2Cl_9^-$

1) DPE
2) MTSP

Hydrolysis

m PVL
18-Crown-6, THF

Scheme 14 Synthesis of poly(IB-*b*-PVL) copolymer by site-transformation

ate potassium salt was successfully used as a macroinitiator for the AROP of PVL in tetrahydrofuran, leading to poly(IB-*b*-PVL) copolymers. The same methodology as mentioned above was applied for the synthesis of poly(PVL-*b*-IB-*b*-PVL) triblock copolymers, except that a difunctional initiator, 5-*tert*-butyl-1,3-bis-(1-chloro-1-methylethyl)-benzene (*t*-BuDiCumCl), was used for the polymerization of IB in the first step. Characterization of the block copolymers by ^{1}H and ^{13}C NMR spectroscopy, high temperature GPC, and DSC showed that this synthetic route provided block copolymers consisting of low glass transition (T_g~−71 °C) PIB segment, and highly crystalline (% crystallinity=55~82%), high melting (T_m~180~225 °C) PPVL segment with predetermined compositions and high structural integrity. Comparison of DSC results with morphological studies using atomic force microscopy, hot stage polarized optical microscopy and small angle X-ray scattering indicated that crystallization of PPVL was constrained to the cylindrical and spherical microdomains preexisting in the melt.

We have further studied the synthesis of novel ABC linear triblock copolymers. Specifically, novel glassy(A)-*b*-rubbery(B)-*b*-crystalline(C) linear triblock copolymers have been investigated, where A block is PαMeSt, B block is rubbery PIB, and C block is crystalline PPVL. The synthesis was accomplished by living cationic sequential block copolymerization to yield living poly(αMeSt-*b*-IB) followed by site-transformation to polymerize PVL [98]. In the first synthetic step, the GPC traces of poly(αMeSt-*b*-IB) copolymers with ω-methoxycarbonyl functional group exhibited bimodal distribution in both RI and UV traces, and the small hump at higher elution volume was attributed to PαMeSt homopolymer. This product was fractionated repeatedly using Hex/ethyl acetate to remove homo PαMeSt and the pure poly(αMeSt-*b*-IB) macroinitiator was then utilized to initiate AROP of PVL to give rise to poly(αMeSt-*b*-IB-*b*-PVL) copolymer.

Complete crossover from living PαMeSt to IB could be achieved by modifying the living PαMeSt chain end with a small amount of *p*ClαMeSt after complete conversion of αMeSt, as presented in Scheme 15. GPC traces of the

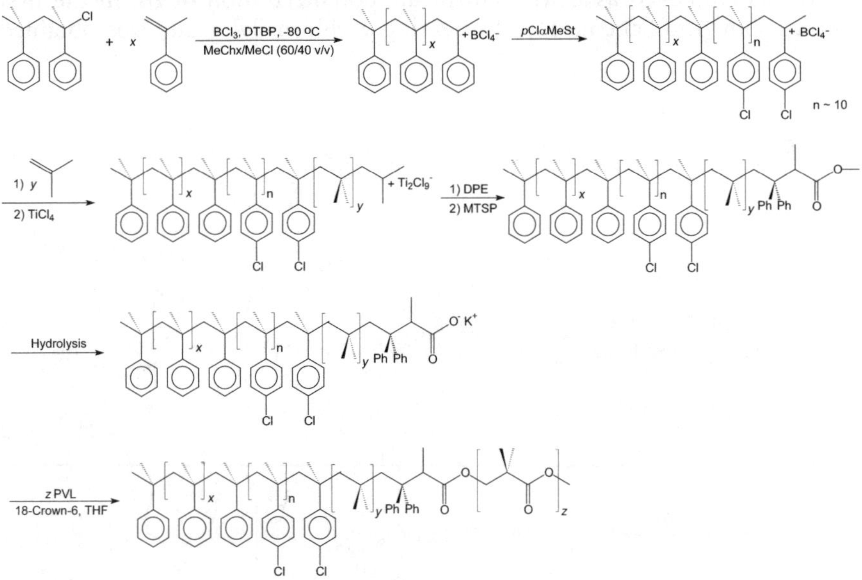

Scheme 15 Synthesis of poly(αMeSt-*b*-*p*ClαMeSt-*b*-IB-*b*-PVL) linear ABCD block copolymer

block copolymer exhibited narrow molecular weight distribution in both RI and UV detection, confirming the synthesis of well-defined diblock copolymers. The poly(αMeSt-*b*-IB) copolymer carrying ω-carboxylate group, obtained from hydrolysis of ω-methoxycarbonyl group of the block copolymer, was used to initiate AROP of PVL in conjunction with 18-crown-6 in THF at 60 °C to give rise to poly(αMeSt-*b*-*p*ClαMeSt-*b*-IB-*b*-PVL) copolymer [98].

5.3
Combination of Cationic and Anionic Polymerization Using Chlorosilyl-Functionalized PIB

The combination of living cationic and living anionic polymerizations provides a unique approach to the synthesis of block copolymers not available by a single method. Coupling of living anionic and cationic polymers is conceptually simple, but few examples have been reported so far. This is most likely due to the different reaction conditions required for living cationic and anionic polymerizations.

The synthesis of chlorosilyl-functional polymers has been of great scientific interest. We have recently invented a new one-pot synthesis of α-chlorosilyl-functional PIBs utilizing novel chlorosilyl-functional initiators for the living cationic polymerization of IB [99, 100]. This was based on a finding that the chlorosilyl head-group of initiators remained intact when $TiCl_4$ was employed as Lewis acid in the polymerization of IB. In the first step as shown in Scheme 16, PIB bearing α-chlorosilyl group was obtained

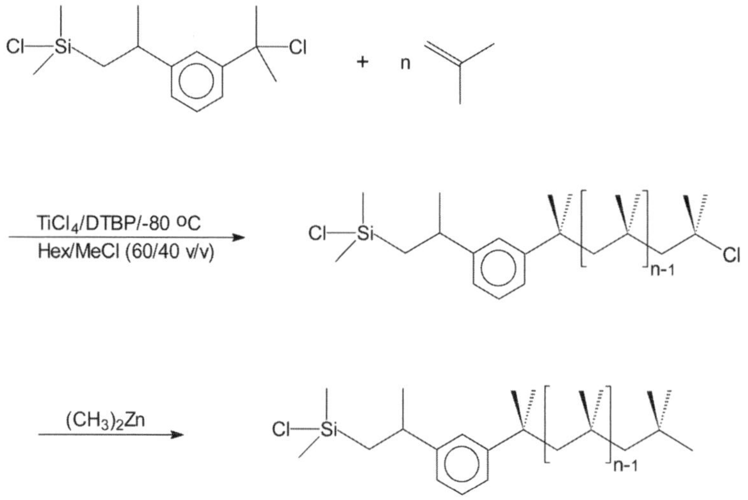

$TiCl_4$/DTBP/-80 °C
Hex/MeCl (60/40 v/v)

$(CH_3)_2Zn$

Scheme 16 Synthesis of α-chlorosilyl, ω-methyl PIB

using the chlorosilyl-functional initiator in conjunction with $TiCl_4$, in the presence of DTBP as proton trap, in Hex/MeCl (60/40 v/v) at −80 °C.

Since the terminal *tert*-chloro group of PIB is thermally and chemically labile, in situ replacement of the *tert*-chloro group with a thermally stable and chemically inert end group is desirable. We have described in situ methylation of living PIB using trimethylaluminum to prepare thermally stable and chemically inert PIB [101]. However, organoaluminum compounds are strong Lewis acids and may induce undesirable side reactions. Quantitative in situ methylation of living PIB has been recently reported using dimethylzinc, which takes place under much milder reaction conditions [102].

Due to the versatility of the chlorosilyl group in chemical reactions, chlorosilyl-functional PIBs, shown in Scheme 16, could be used as intermediates for block copolymers in coupling reactions with living anionic polymers. Research in this area is ongoing in our laboratory.

6
Conclusions

Living cationic polymerization and the combination of living polymerization techniques have been proven to be powerful tools in the design and synthesis of PIB-based block copolymers. These methods provide well-defined and compositionally homogeneous linear and star-branched materials. Block copolymers combining immiscible plastic-rubbery, crystalline-rubbery, and hydrophobic-hydrophilic block segments are valuable model materials for structure-property investigation, and potentially useful materials as elastomers, impact modifiers, compatibilizers, adhesives, etc. Growing industrial interest especially in thermoplastic elastomers and materials for biomedical applications encourages further research in this area.

Acknowledgments Financial support from National Science Foundation (DMR-9502777 and DMR-9806418), the Exxon Chemical Co. and Dow Corning Corp. is gratefully acknowledged.

References

1. Miyamoto M, Sawamoto M, Higashimura T (1984) Macromolecules 17:265
2. Faust R, Kennedy JP (1986) Polym Bull 15:317
3. Faust R, Kennedy JP (1988) Polym Bull 19:21
4. Kennedy JP (1996) In: Holden G, Legge, NR, Quirk RP, Schroeder HE (eds) Thermoplatic elastomers, 2nd edn. Hanser Publishers, New York, chap 13
5. Holden G (2000) Understanding thermoplastic elastomers, Hanser Publishers, Munich
6. Mayr H (1996) In: Matyjaszewski K (ed) Cationic polymerization. Mechanism, synthesis, and application. New York, p 125
7. Fodor Zs, Gyor M, Wang HC, Faust R (1993) J Macromol Sci Pure Appl Chem A30:349
8. Faust R, Cao X (1999) Polym Prepr (Am Chem Soc, Div Polym Chem) 40:1039
9. Cao X, Faust R (1999) Macromolecules 32:5487
10. Cao X, Faust R (1998) Polym Prepr (Am Chem Soc, Div Polym Chem) 39:496
11. Shim JS, Asthana S, Omura N, Kennedy JP (1998) J Polym Sci Polym Chem 36:2997
12. Storey RF, A.Shoemake K (1996) Polym Prepr (Am Chem Soc, Div Polym Chem) 37:321
13. Kim MS, Faust R (2003) Macromolecules (submitted)
14. Higashimura T, Kamigaito M, Kato M, Hasebe T, Sawamoto M (1993) Macromolecules 26:2670
15. Matyjaszewski K, Bon A, Lin ChH, Xiang JS (1993) Polym Prepr (Am Chem Soc, Div Polym Chem) 34:487
16. Li D, Hadjikyriacou S, Faust R (1996) Macromolecules 29:6061
17. Li D, Hadjikyriacou S, Faust R (1996) Polym Prepr (Am Chem Soc, Div Polym Chem) 37:803
18. Fodor Zs, Faust R (1998) J Macromol Sci Pure Appl Chem A35:375
19. Kwon Y, Cao X, Faust R (1998) Polym Prepr (Am Chem Soc, Div Polym Chem) 39:494
20. Kwon Y, Cao X, Faust R (1999) Macromolecules 32:6963
21. Kwon Y, Cao X, Faust R (1999) Polym Prepr (Am Chem Soc, Div Polym Chem) 40:1035
22. Cao X, Sipos L, Faust R (2000) Polym Bull 45:121

23. Sipos L, Cao X, Faust R (2001) Macromolecules 34:456
24. Thunogae Y, Kennedy JP (1994) J Polym Sci Polym Chem A32:403
25. Lenz RW (1975) J Macromol Sci Chem A9:945
26. Pernecker T, Kennedy JP, Ivan B (1992) Macromolecules 25:1642
27. Pernecker T, Kennedy JP, Ivan B (1992) Polym Bull 29:27
28. Lubnin AV, Kennedy JP (1993) J Polym Sci Polym Chem 31:2825
29. Faust R (1999) Polym Prepr (Am Chem Soc, Div Polym Chem) 40:960
30. Bae YC, Fodor Zs, Faust R (1997) ACS Symp Ser 665:168
31. Charleux B, Moreau M, Vairon JP, Hadjikyriacou S, Faust R (1998) Macromol Symp 132:25
32. Schlaad H, Erentova K, Faust R, Charleux B, Moreau M, Vairon JP, Mayr H (1998) Polym Prepr (Am Chem Soc, Div Polym Chem) 39:490
33. Schlaad H, Erentova K, Faust R, Charleux B, Moreau M, Vairon JP, Mayr H (1998) Macromolecules 31:8058
34. Schlaad H, Kwon Y, Faust R, Mayr H (2000) Macromolecules 33:743
35. Canale PL, Faust R (1998) Polym Prepr (Am Chem Soc, Div Polym Chem) 39:400
36. Canale PL, Faust R (1999) Macromolecules 32:2883
37. Li D, Faust R (1995) Macromolecules 28:1383
38. Fodor Zs, Faust R (1994) J Macromol Sci Pure Appl Chem A31:1985
39. Hadjikyriacou S, Faust R (1996) Polym Prepr (Am Chem Soc, Div Polym Chem) 37:345
40. Hadjikyriacou S, Faust R (1996) Macromolecules 29:5261
41. Hadjikyriacou S, Faust R (1995) Macromolecules 28:7893
42. Hadjikyriacou S, Faust R (1995) Polym Prepr (Am Chem Soc, Div Polym Chem) 36:174
43. Hadjikyriacou S, Faust R (1997) US Patent Application 08/967,002 filed November 10
44. Hadjikyriacou S, Faust R (1998) Polym Prepr (Am Chem Soc, Div Polym Chem) 39:398
45. Kwon Y, Hadjikyriacou S, Faust R, Cabrit P, Moreau M, Charleux B, Vairon JP (1999) Polym Prepr (Am Chem Soc, Div Polym Chem) 40:681
46. Hadjikyriacou S, Faust R (1999) Macromolecules 32:6393
47. Kennedy JP (1996) In: Holden G, Legge NR, Quirk RP, Schröder HE (eds) Thermoplastic elastomers, 2nd edn. Hanser Publishers, Munich, p 365
48. Kaszas G, Puskas J, Kennedy JP, Hager WG (1991) J Polym Sci Polym Chem A29:427
49. Everland H, Kops J, Nielson A, Iván B (1993) Polym Bull 31:159
50. Storey RF, Chisholm BJ, Choate KR (1994) J Macromol Sci Pure Appl Chem A31:969
51. Gyor M, Fodor Zs, Wang HC, Faust R (1993) Polym Prepr (Am Chem Soc, Div Polym Chem) 34:562
52. Gyor M, Fodor Zs, Wang HC, Faust R (1994) J Macromol Sci Pure Appl Chem A31:2055
53. Fodor Zs, Faust R (1995) Polym Prepr (Am Chem Soc, Div Polym Chem) 36:176
54. Fodor Zs, Faust R (1996) J Macromol Sci Pure Appl Chem A33:305
55. Storey RF, Baugh DW, Choate KR (1999) Polymer 40:3083
56. Gyor M, Fodor Zs, Wang HC, Faust R (1993) Polym Prepr (Am Chem Soc, Div Polym Chem) 34:562
57. Kennedy JP, Kurian J (1990) J Polym Sci Polym Chem 28:3725; (1990) Polym Mater Sci Eng 63:371
58. Kennedy JP, Keszler B, Tsunogae Y, Midha S (1991) Polym Prepr (Am Chem Soc, Div Polym Chem) 32:310
59. Kennedy JP, Meguriya N, Keszler B (1991) Macromolecules 24:6572
60. Tsunogae Y, Kennedy JP (1993) J Macromol Sci Pure Appl Chem A30:269
61. Everland H, Koops J, Nielsen A, Ivan B (1993) Polym Bull 31:159
62. Tsunogae Y, Kennedy JP (1994) J Polym Sci Polym Chem 32:403
63. Li D, Faust R (1995) Macromolecules 28:4893
64. Fodor Zs, Faust R (1995) J Macromol Sci Pure Appl Chem A32:575
65. Fukui H, Sawamoto M, Higashimura T (1993) J Polym Sci Polym Chem 31:1531
66. Fukui H, Sawamoto M, Higashimura T (1993) Macromolecules 26:7315
67. Fukui H, Sawamoto M, Higashimura T (1996) Macromolecules 29:1862
68. Bae YC, Hadjikyriacou S, Schlaad H, Faust R (1999) In: Puskas JE (ed) Ionic polymerization and related processes. Kluwer Academic Publishers, Dordrecht, Netherlands

69. Bae YC, Fodor Zs, Faust R (1997) Macromolecules 30:198
70. Bae YC, Coca S, Canale PL, Faust R (1996) Polym Prepr (Am Chem Soc, Div Polym Chem) 37:369
71. Gandini A (1992) In: Aggarwal SL, Russo SL (eds) Comprehensive polymer science, supplement 1. Pergamon, Oxford, UK, p 527
72. Charleux B, Faust R (1998) Adv Polym Sci 142:1
73. Shohi H, Sawamoto M, Higashimura T (1991) Polym Bull 25:529
74. Kaszas G, Puskas JE, Kennedy JP, Hager WG (1991) J Polym Sci Polym Chem 29:427
75. Storey RF, Chisholm BJ, Lee Y (1993) Polymer 34:4330
76. Fodor Zs, Faust R (1995) J Macromol Sci Pure Appl Chem A32:575
77. Jacob S, Majoros I, Kennedy JP (1997) Polym Mater Sci Eng 77:185; (1998) Rubber Chem Technol 71:708
78. (a) Jacob S, Kennedy JP (1998) Polym Bull 41:167; (b) Jacob S, Majoros I, Kennedy JP (1998) Polym Prepr (Am Chem Soc, Div Polym Chem) 39:198
79. Puskas JE, Pattern W, Wetmore PM, Krukonis V (1999) Polym Mater Sci Eng 80:429; (1999) Rubber Chem Technol 72:559
80. Brister LB, Puskas JE, Tzaras E (1999) Polym Prepr (Am Chem Soc, Div Polym Chem) 40:141
81. (a) Storey RF, Shoemake KA (1996) Polym Prepr (Am Chem Soc, Div Polym Chem) 37:321; (b) Storey RF, Showmake KA (1999) J Polym Sci Polym Chem 37:1629; (c) Asthana S, Majoros I, Kennedy JP (1998) Rubber Chem Technol 71:949; (d) Asthana S, Kennedy JP (1999), J Polym Sci Polym Chem 37:2235
82. (a) Shim JS, Asthana S, Omura N, Kennedy JP (1998) J Polym Sci Polym Chem 36:2997; (b) Shim JS, Kennedy JP (1999) J Polym Sci Polym Chem 37:815
83. Shim JS, Kennedy JP (2000) J Polym Sci Polym Chem 38:279
84. Quirk RP, Yoo T, Lee Y, Kim J, Lee B (2000) Adv Polym Sci 153:67
85. Bae YC, Faust R (1998) Macromolecules 31:2480
86. Bae YC, Faust R (1998) Macromolecules 31:9379
87. Yun JP, Hadjikyriacou S, Faust R (1999) Polym Prepr (Am Chem Soc, Div Polym Chem) 40:1041
88. Takács A, Faust R (1995) Macromolecules 28:7266
89. (a) Kennedy JP, Price JL (1991) Polym Mater Sci Eng 64:40; (b) Kitayama T, Nishiura T, Hatada K (1991) Polym Bull 26:513; (c) Kennedy JP, Price JL, Koshimura K (1991) Macromolecules 24:6567; (e) Nishiura T, Kitayama T, Hatada K (1992) Polym Bull 27:615
90. Feldhusen J, Iván B, Müller AHE (1997) Macromolecules 30:6989
91. (a) Feldhusen J, Iván B, Müller AHE (1998) Macromolecules 31:4483; (b) Feldthusen J, Iván B, Muller AHE (1998) ACS Symp Ser 704:121
92. Sipos L, Zsuga M, Deák G (1995) Macromol Rapid Commun 16:935
93. Gorda KR, Peiffer DG, Chung TC, Berluche E (1990) Polym Commun 31:286
94. (a) Storey RF, Sherman JW, Brister LB (2000) Polym Prep. (Am Chem Soc, Div Polym Chem) 41:690; (b) Storey RF, Brister LB, Sherman JW (2001) J Macromol Sci Pure Appl Chem A38:107
95. Kim MS, Faust R (2003) Polym Bull (in press)
96. Kwon Y, Faust R (2000) Polym Prepr (Am Chem Soc, Div Polym Chem) 41:1597
97. Kwon Y, Faust R, Chen CX, Thomas EL (2003) Polym Mater Sci Eng 84 (submitted)
98. Kwon Y, Kim MS, Faust R (2001) Polym Prep (Am Chem Soc, Div Polym Chem) 42:483
99. Kim IJ, Faust R (2000) Polym Prep (Am Chem Soc, Div Polym Chem) 41:1309
100. Faust R, Hadjikyriacou S, Roy AK, Suzuki T (2000) PCT Int Appl p 24
101. Takacs A, Faust R (1996) J Macromol Sci Pure Appl Chem A33:117
102. Bae YC, Kim IJ, Faust R (2000) Polym Bull 44:453

Editor: Oskar Nuyken
Received: May 2002

Author Index Volumes 101–167

Author Index Volumes 1-100 see Volume 100

Subject Index

Printing: Saladruck, Berlin
Binding: Stein+Lehmann, Berlin